Advance Praise for

Ecoportraiture: The Art of Research When Nature Matters

"This book recognises that we need 'new ways of witnessing, interpreting, mulling over, reflecting on, listening to and dialoguing with the human and the more-than-human'. This is not for the purpose of novel research to serve an education industry, but because we need better ethical understandings if we are to seriously address our relationships with the more-than-human world and avert future global catastrophes. *Ecoportraiture* not only offers improved intellectual knowings, but is also a portal to much needed expanded existential understandings. As befitting of profound and beautiful ideas, it is also beautifully written. It's a book that needs to be read by anyone who is serious about research and anyone who is serious about education."

—Dylan Adams, Cardiff Metropolitan University

"What a wonderous and powerful way to approach research in and with the natural world! To be at once aware of 'what is already at work without me' and dedicated to listening ethically through engaging ecoportraiture as method seems exactly right to me. I wish I had this book when I was teaching qual methods."

—Rebecca Martusewicz, Emerita, Eastern Michigan University

Ecoportraiture

Constance Russell and Justin Dillon
General Editors

Vol. 16

The [Re]Thinking Environmental Education series
is part of the Peter Lang Education list.
Every volume is peer reviewed and meets
the highest quality standards for content and production.

PETER LANG
New York • Bern • Berlin
Brussels • Vienna • Oxford • Warsaw

Ecoportraiture

The Art of Research When Nature Matters

Edited by
Sean Blenkinsop, Mark Fettes,
and Laura Piersol

PETER LANG
New York • Bern • Berlin
Brussels • Vienna • Oxford • Warsaw

Library of Congress Cataloging-in-Publication Data

Names: Blenkinsop, Sean, editor. | Fettes, Mark, editor. | Piersol, Laura, editor.
Title: Ecoportraiture: the art of research when nature matters / edited by Sean Blenkinsop, Mark Fettes, and Laura Piersol.
Description: New York: Peter Lang, 2022.
Series: [Re]thinking environmental education; Vol. 16 | ISSN 1949-0747
Includes bibliographical references and index.
Identifiers: LCCN 2021040927 (print) | LCCN 2021040928 (ebook)
ISBN 978-1-4331-9256-2 (hardback)
ISBN 978-1-4331-9289-0 (ebook pdf) | ISBN 978-1-4331-9290-6 (epub)
Subjects: LCSH: Environmental education. | Education—Environmental aspects—Research. | Place-based education. | Nature study. | Outdoor education.
Classification: LCC LC239 E36 2022 (print) | LCC LC239 (ebook) | DDC 371.3/84—dc23
LC record available athttps://lccn.loc.gov/2021040927
LC ebook record available at https://lccn.loc.gov/2021040928
DOI 10.3726/b19182

Bibliographic information published by **Die Deutsche Nationalbibliothek**.
Die Deutsche Nationalbibliothek lists this publication in the "Deutsche Nationalbibliografie"; detailed bibliographic data are available on the Internet at http://dnb.d-nb.de/.

80 Broad Street, 5th floor, New York, NY 10004
www.peterlang.com

Table of Contents

List of Illustrations

List of Tables

Foreword: Wanting to Be Felt Seen

David W. Jardine

Photo: D. Jardine

Portraying has similar etymological ancestries and suppressed memories as "education." Drawing forth, drawing out, tracing. And, I'd add from hermeneutics, highlighting, exaggerating, even the tricky lies of tricksters like Hermes bent on telling the truth, sort of.

But it was the Raven that drew *me* out, here, so who is the portrayer and who portrayed is, I think, the sense in which portraiture is always about relationships, and, I'd add, about precisely the ecological *ecstasis* that comes from this very switcheroo. Not just that "I wanted the subject to feel seen" (Lawrence-Lightfoot & Davis, 1997, p. 4; see Fettes, Blenkinsop & Piersol, Chapter 1, this volume, p. 7), but *I wanted to be felt seen*. Being regarding and regarded, sensing the line traced on my face by that glance and the soft downbeak of rictal feathers.

Snap.

Shot.

And ending up with tailfeathers clustered like thunderclouds or a Chinook Arch. My tailfeathers sparkled up with spine words at the sight of him there. Lesley Tait sending me a story of Raven's consorting with lightning. I daresay "of course" even though I'd never heard that story before. We're trying to write something about it. Stayed tuned (that perhaps the best metaphor for this whole book).

Sketchy, all this. Look at that Raven. My measure has been taken and, in putting that photograph there, I must now attest to how I, too, am portrayed by that Raven's glare and pause and beauty. Portrayed in his portrait that I took, as will anyone be who knows how to look at a photograph and let it work.

That grey dew-chill on grass blades (see Piersol in Chapter 4, this volume) portrays me, clear as day, as the one who knows something of that chill and whose bones are more full than ever of its reminder as I creak over 70. I've been around it and now adore its being noted enough to get me to write out this very line of words in praise, in recognition of being recognized.

I've been spotted. Just like the kid-up-a-tree (Chapter 4) aches my heart over an old book cover (Jardine, 1998) now 23 years old, named after a paper that is itself now 30 years old (Jardine, 1990):

Cover design by N. Reuter (Jardine, 1998).

That very boy-of-then having just warned me of an "atrociously long sentence" in a piece I wrote in late November. I don't feel too bad about this dribbling trail of self-citation, because it might give readers a sense of just how many trees I've pissed on, and for how long, and therefore how deeply and longstandingly my heart aches and leaps with joy thumbing this book.

A "critical, beautiful risk" (Datura in Chapter 6, this volume), full of "fire and clinical coldness" (Joan Miró, cited by de Grandpré, Chapter 5, this volume) – ways to do research, full of great warrant and great, longstanding, shared and contested ancestries, shared and contested voices whose share and contest is a necessary part of the portraits themselves. These matters are not singular. These stories have no moral from which one might deviate, or to which one might try to slavishly adhere. Such singularity is precisely the terrible consort of exactly the severances this book is wont to overcome. And, to be blunt, such singularity is, more often than not, precisely what so many "research methods" wittingly or unwittingly seek. So many are querulous over beauty, suggestion, ambiguity, participation, refusal, multivocity, repetition, fruity language, hesitancy. But here's a methodological warrant for you, originally from Aristotle: *mensuratio ad rem* (cited in and discussed in Gadamer, 1989, p. 261). Good work finds its measure, not in the method of its approach, but in the object of its attention and devotion, and it shapes its own approach, its own portrait of that object in that proper measure. We can, with long practice and great care, "entrust ourselves to what we are investigating to guide us safely in the quest" (p. 378).

In the gaze of that Raven, hesitancy is a good way to speak truly of it. To let it be uncaptured by the words I use, because that is what it *is*. To use words that hand readers over to that uncapture.

This might sound strange, but relax, it is very simple and straight forward. Blazing Red Maple Leaves (from Timmerman, Chapter 2, this volume), a portrait of a place I was raised that was not taken anywhere near where I was raised, a portrait of my no longer living under their penumbras of seasonal Lake Ontario turning, a portrait of loss gained in the looking, of blood waning painted up in the sky. Memory. Absence: there are no Maples nearby. It's a smell I know I'd recognize. A kinship with more stories in it than any one heart can possibly tell. We don't need a method to tell them all or to glean their essence. We need a method that will give readers the fulsome experience of that very abundance, that very beyond-one-heart.

Including bird-hearts. Including the shrills of in-sections:

> How things smell, the racket of leaves turning on their stems, how my breath pulls this humid air, how birds songs combine, the familiar directions of sudden thundery winds, the rising insect drills of cicada tree buzzes that I remember so intimately, so immediately, that when they sound, it feels as if this place itself has remembered what I have forgotten, as if my own memory, my own raising, some of my own life, is stored up in these trees for safe keeping. Cicadas become archaic storytellers telling me, like all good storytellers, of the life I'd forgotten I'd lived, of deep, fleshy, familial relations that worm their ways out of my belly and breath into these soils, these smells, this air. And I'm left shocked that they know so much, that they remember so well, and that they can be so perfectly articulate. (Jardine, 2016, pp. 83–4)

And being left knowing full well that, if or when I meet Nora Timmerman, we have this lovely thing in common, our difference, and how that difference illuminates both sides of the relationship, illuminates the underside of those leaves, penumbras umbrellaing the light. How it shows Red Maple Leaves as "hospitable" (Fettes, Blenkinsop & Piersol, p. 3) enough to sustain, cultivate and care for such meeting, without dissolving into identity or difference, neither of which are adequate to what this book explores.

"Reconnecting with nature and place" must mean, of course, reconnecting with my own nature and place and reconnecting to all the distracting and exhausting ways that that reconnecting is so easily frazzled, so easily mis-named or misappropriated. And it must also mean, I suggest, reconnection with the nature and topography of *any and all* of the things we teach in schools, inside the walls and out. *All of it.* Our whole pedagogical countenance must be ecological. How could it be otherwise than a way to experience, explore, understand, adore *everything.* If we leave some of the ancestries of knowledge behind inside the school walls while we while outside in gracious acts of portraiture, we simply reproduce the very severances we were out there to heal. If we leave some of them behind, we, however unintentionally and inadvertently, fall for precisely the root-severance in the very act of trying to no longer fall for it. Citing Sara Lawrence-Lightfoot and Jessica Hoffman Davis (1997), the "inspiration for the studies collected here" (Fettes, Blenkinsop, & Piersol, p. 3), our "search for goodness" must be itself be as good as it can be.

There is nothing left over. There is no elsewhere elsewhither else-whence. The inside and the outside of school walls is a hallucination and "getting out," by itself, ain't good enough. And, as I've witnessed first hand and as this book so delightfully and deliberately describes, I can start right here: to the extent that I'm sucking in this air, heavy with wood split and dried and stacked and now December-carried, I'm already portrayed – my body long since etched and sketched right at the cellular level – by the trees that cast up this oxygen and give this wood over that warms my insides.

But that creepy, lovely sense of intimacy and beholding and beholdening (looking at a portrait is not just beholding it but becoming beholden to it) must apply to the uproariousness of laughter with a Grade Two class as we searched out the origins of the English names of the days of the week and found Norwegian Gods hiding in plain sight. Talk about "radical inclusion and reflexivity" (Fettes, Blenkinsop & Piersol, p. 3) as children in the class familiar with other tongues came to school with names and counterpoints, faultlines and interdependencies, and other Monday-morning lunacies. And just like the oxygens I've drawn for every day of my living, I used those English names *for decades and decades* as if there were no debts to understand, no places to explore, no ancestries to parse and parry, no stories already told about me without a glimmer of aware, no especially familiar other names that might put my own untold stories in highlight. These suddenly experienced severances were quite literally *humiliating* to find out and then take on in slow walks, week by week, through the myriad days, glad of the different licking tongues children brought with them. Glad to have lived through an instance where those indigenous to English, like me, without a drop of English blood, were as unaware of these English names as anyone else. Just living in the place every morning as the days scroll by doesn't mean that you are *living in the place*, even though you're living in the place. Sylvie de Grandpré (p. 97). notes "how layers of colonialism have silenced the natural world, and how colonial conceptualizations of the natural world maintain the status quo." It is good to remember, as this book and her chapter unearth, how much my own withery Whiteness doesn't remember much *about itself* – its own words, its own past, its own actions – let alone this place, old names for the ways of this place and those of this place.

There. Right there is the small miracle of this book, because that ear-popping search, that eye-opening moment, is "the path and the goal" (Gadamer, 1989, p. 180) of these explorations, these portraitures, "iterative and generative" as Keena (Chapter 3, this volume, p. 44) both cites and portrays so well. Her final sentence: "the stories of many children in context and relationship became the rich texture and detail of a specific place [that] illuminate the place itself" (p. 56). In this way of working laid out in these pages, the stories are becoming of the very place of which they tell, iterative and generative. Let me push it further – the stories *are* as much a part of the rich texture and detail of a specific place as the textures and details of the specific place itself. It must be said this way. Otherwise, severance re-ensues all over again.

"Addressing the structuring of dominant cultures' speech is one of the ways towards decolonization" (de Grandpré, p. 96), and this includes – this *must* include, of course – the de-colonization of the colonizers as well, a shaking out of the language we've inherited or had imposed or presumed upon us from these winding legacies. So, I'll end with a qualm – small but rather pernicious.

What is laid out in this book does not, in any way "capture" the "essence" (Lawrence-Lightfoot & Davis, 1997, p. 4) of the subjects it portrays. Essence – that word, in the phenomenological tradition, is mired in hallucinations of fixity, singularity and finality and "once and for all" (Husserl, 1970, 135), however much this might be unintended by its usage. It was wielded as a way to judge the proximity or distance of this or that case from the essentials of it. It was a way to sever off essence from accident, where the particular story, photograph, lament, joyous outburst, are – Husserl actually used this terminology – "purified" of accidental features, such that we need not hear the particular story again because now we know what its essence is. In fact, *the next story* – e.g., the next classroom case of portraiture, the next instance of ecological horror, ecological joy, or colonial creep – becomes, under the auspices of "essences," *less and less important to heed* because, by having already sussed the essence of such matters, the next cases are less and less revelatory except for accidental details. I wanted to make sure I stopped over this because the eradication of the need to hear the next case, the next story, is the *goal* of many research methodologies. Portraiture, like hermeneutics and the other threads that these various authors summon, *works precisely against this very eradication.* "Essence" is the core of colonialism and the bane of portraiture's lovely grace.

Humming here together, writing, reading, there is no essence had, wanted or sought, just the interrelations, the gathering, the movement of thought braiding itself up over the foothills to the West, breaths, voices, each new one making all the others more dear, more legible and true.

It's just why this book is so good to see, and will be so nice to hold and linger over by hand. So, in thanks:

Words written after a symphony of
Tendon-plucking oldbody hum and sunwalking purrs of quiet and
Rests and dogchorouses. Wood carries.
Breaths of ice air. Raven gurglings
And the dicey curves of greygreen dew on
Greygreen grass.

It's bonechilly.
I pull a loose collar of skin up around my neck
For warmth, then
Sit.
To write.

dj

Bragg Creek, AB.
Signed off, December 3, 2020, 9:35 AM, MST
17 days 'til Sol Stops.
Again.
"Oh, please come back around" sung
To the tune of Xmas carols.

References

Gadamer, H. G. (1989). *Truth and method.* New York, NY: Continuum Press.

Husserl, E. (1970b). *The crisis of European science and transcendental phenomenology.* Chicago, IL: Northwestern University Press.

Jardine, D. (1990). "To dwell with a boundless heart": On the integrated curriculum and the recovery of the Earth. *Journal of Curriculum and Supervision*, 5(2), 107–119.
Jardine, D. (1998). *"To dwell with a boundless heart": On curriculum theory, hermeneutics and the ecological imagination.* New York, NY: Peter Lang.
Jardine, D. (2016). Birding lessons and the teachings of cicadas. In D. Jardine (Ed.), *In praise of radiant beings: A retrospective path through education, Buddhism and ecology* (pp. 83–96). Charlotte, NC: Information Age Publishing.
Lawrence-Lightfoot, S., & Davis, J. H. (1997). *The art and science of portraiture.* San Francisco: Jossey-Bass.

1. *Ecoportraiture: Researching in Resonance with the More-Than-Human*

MARK FETTES, SEAN BLENKINSOP, & LAURA PIERSOL

Starting with Some Good Questions

The ecological crisis is, among other things, a failure of education to connect children and adults more closely to the life of the world. We are, of course, deeply and inextricably part of that life, whether we are aware of it or not. But how we understand our responsibilities, how we conduct ourselves, how we treat our more-than-human kin and the places we share with them – all of these depend, crucially, on how we are educated, both informally in the contexts of family, community and the broader culture, and formally, in schools.

Formal education is in fact a late, and largely destructive, intruder on the scene. Over tens of millennia, Indigenous peoples the world over have developed sophisticated practices of kinship and reciprocity with the myriad species who live alongside them; such practices are cultivated, not in classrooms or other human-dominated environments, but in direct contact with the living world. It is the worldwide spread of compulsory, industrial-style formal education over the past century and a half that has marginalized those ancient traditions. Many children now spend their years of schooling essentially cut off from all forms of intentional relationality with other species, including the birds in the trees outside the school, the trees themselves, the insects in their bark, the fungi in the soil, and beyond all these things the wild nature of the place itself, as it exists and unfolds on its own terms and for its own sake. And this severance comes at a price – a heavy one, in both human and ecological terms.

Awareness of that price has grown in recent decades. As a consequence, it is more and more common to encounter local initiatives aimed at reconnecting formal education with nature and place, including forest schools (Knight,

2013; Sobel, 2016), outdoor classrooms and eco-learning centers (Renshaw & Tooth, 2017; Waite, 2011), and programs in place-based and nature-based learning (Anderson, 2017; Smith & Sobel, 2010). Indigenous educators, too, have been placing increasing emphasis on learning on the land, both in the context of the revitalization of languages and cultures impacted by colonialism and as a pathway towards reconciliation and healing for individuals and for society as a whole (McCoy, Tuck, & McKenzie, 2017; Styres, 2017).

What changes, as learning moves from classrooms and built environments into parks and forests and along watercourses and shorelines? What changes, as weather and the succession of seasons start to influence where and how learning happens? What changes, as the focus of attention shifts from patterns of text to patterns of texture, from the echoes of the hallway to the birdsong of the grove, from the smells of chalk and disinfectant to the fragrance of soil? What changes, when it is not just humans doing the teaching? Implicit in these moves is a process of gradual transformation for students, teachers and families, a reweaving of their relationships with land and place that shifts their understandings of themselves as well.

As educational researchers, we are confronted with the challenge of finding a methodology of inquiry that is resonant with these relational shifts. One can, of course, employ well-established qualitative approaches (case study, ethnography, grounded theory, and the like) to find out how the people involved make sense of their experiences and learning over time. The problem is that such methodologies risk re-inscribing a human-centered conception of this process of transformation, as if it were simply a matter of people choosing to teach and learn differently. It is, of course, that; people's choices and actions and understandings are integral to how this all unfolds. But if this unfolding entails coming to relate to the more-than-human world (Abram, 1996) not merely as a backdrop to human life, but as kin, interlocutor, teacher and guide, then methodologies are needed that are accountable to this shift.

In part, this is a question about storytelling. Stories, as Thomas King observes, "are wondrous things. And they are dangerous" (2003, p. 9). In what stories focus on and what they leave out, in how they are told and who does the telling and for what purpose, they convey cultural attitudes, values and priorities. The stories Western civilization has been telling about the more-than-human world have encouraged us to see it as subservient to our needs; the stories it tells about human beings encourage us to see connections to place and land as relics of the past, vestiges of the time before we grew up and became modern (Stibbe, 2015). There are, of course, important countercurrents in Western culture itself, ranging from the 19th-century Romantics who championed poetic expression and natural beauty in opposition to the Industrial Revolution (Rigby, 2014), through such touchstones of modern environmentalism as Muir, Carson and Leopold (Payne, 1996), to contemporary poets and essayists (e.g. Berry, 2012; Dillard, 2015; Oliver, 2013; Wagoner, 1999) who explore the meaning of the more-than-human world as it presents itself to individual consciousness. A variety of non-Western traditions (Selin, 2013) have contributed to this flow of poetic or narrative or philosophic inquiry that calls us towards more respectful relationship with the places and the planet where we live. "Want a different ethic? Tell a different story," says King (2003, p. 164), and that is what this literature has done and continues to do, in a variety of voices and ways (Garrard, 2012, 2014; Zapf, 2016).

And yet this is not really a *research* literature, still less a literature that focuses on shared educational spaces and processes as they unfold over time. One can of course draw educational implications from these works, yet education itself, that uncertain and emergent enterprise filled with a diversity of voices and experiences, human and more-than-human, is rarely portrayed. An outstanding (though partial) exception is Robin Wall Kimmerer's *Braiding Sweetgrass* (2013), in which several chapters lovingly depict her work with university students, and many more consider plants as teachers in the broader context of American culture. A way of posing our central question might thus be the following: can the depth of educational attention and insight that Kimmerer displays be cultivated within a research tradition? What would such a methodology look like? And how might the products of such research inform the work of bringing more such spaces and processes into being?

The "narrative turn" and other developments in the social sciences over the past 30–40 years have brought us to a place where such questions can be asked more creatively and productively (Gough, 2013; Stevenson, Dillon, Wals, & Brody, 2013). In its initial phases, research on "environmental education," as the field became known in the 1970s and 1980s, was oriented to "fostering responsible environmental behaviour"; it was seen as applied research directed to achieving well-defined outcomes, typically framed in terms of individual awareness, understanding, values, skills and action. Gradually, the scope of the field has expanded to include critical, feminist, post-modern, post-human, new materialist, Indigenous and other approaches that raise vital questions about whose voices and viewpoints are represented in research (Gough, 2013). In a variety of ways, these bodies of scholarship are also expanding our conceptions of how language and other mediational means can be used, in a research context, to recognize, evoke, provoke and interrogate our embeddedness in the more-than-human world.

Portraiture, the North American research tradition that served as inspiration for the studies collected here, is founded, in the words of its leading exponent Sara Lawrence-Lightfoot, on "productive and benign relationships" among the participants in the research, and oriented towards "a search for goodness" in the research setting (Lawrence-Lightfoot & Davis, 1997). Such language can sound overtly *uncritical* in the context of the fierce ongoing struggles over meaning and representation that characterize educational research today (Reid & Payne, 2013). Yet, as this and the following chapters may help to make clear, these core principles of portraiture are hospitable to radical inclusion and reflexivity; they reflect an underlying ethical stance of *openness* to the complex intertwining of subjectivity, agency and interdependence inherent in ecological relationship, and to the shadows and faultlines that inevitably form part of the whole. Implicit in portraiture is a research ethics that seems well adapted to a time of ecological crisis, when human societies seem so far removed from a state of flourishing-in-place, yet myriad local initiatives are trying to heal our relationships with the more-than-human world.

Our goal in this book, then, is to share, and think about, some ways that a number of eco-educators, typically in the course of their doctoral research, have been drawn to take up and adapt the methodology of portraiture so that it has come to resonate with their own values, concerns and ecological relationality. Each researcher has done so in their own way, and yet, reading their accounts, one is struck by the frequent recurrence of shared principles, values, concerns, struggles, and dilemmas. Together, these narratives map the evolving terrain of *eco*-portraiture. Some of its main features are sketched out in this introductory chapter; others we have left you to discover on your own, reading through the chapters that follow; while in the Epilogue, we offer some forward-looking thoughts and questions. Our intent is to offer not a how-to manual, but a stimulus to reflection and exploration, in the hope that this may become a broad and varied field of inquiry as more researchers in more places take up this work, challenging but also full of joy.

So: let us begin at the beginning. . .

The Origins of Portraiture

The Art and Science of Portraiture was published in 1997, as a joint project of Sara Lawrence-Lightfoot and Jessica Hoffman Davis, a professor and a lecturer in the Harvard Graduate School of Education with disciplinary backgrounds in sociology and the arts, respectively. Lawrence-Lightfoot describes its subject as "a method of qualitative research that blurs the boundaries of aesthetics and empiricism in an effort to capture the complexity, dynamics, and subtlety of human experience and organizational life" (1997, p. xv). The book consists of alternating chapters written by each author, and is illustrated with drawings by children and adults.

By the time the book was written, Lawrence-Lightfoot had been exploring the ideas and methods of portraiture for some 15 years, beginning with *The Good High School* (1983). She was reacting not only against the traditional objectifying gaze of Western social science, but also against "the general tendency of social scientists to focus their investigations on pathology and disease rather than on health and resilience" (1997, p. 8). She also wanted to write in ways "that will seduce the readers into thinking more deeply about issues that concern them" (1997, p. 10). To accomplish this, she saw the need to focus on the particulars of individual voices and narratives, but always *in context* and with explicit recognition of the "self of the portraitist... as an instruments of inquiry, an eye on perspective-taking, an ear that discerns nuances, and a voice that speaks and offers insights" (1997, p. 13). Thus, she summarizes:

> ...portraiture is a method framed by the traditions and values of the phenomenological paradigm, sharing many of the techniques, standards, and goals of ethnography. But it pushes against the constraints of those traditions and practices in its explicit effort to combine empirical and aesthetic description, in its focus on the convergence of narrative and analysis, in its goal of speaking to broader audiences beyond the academy (thus linking inquiry to public discourse and social transformation), in its standard of authenticity rather than reliability and validity (the traditional standards of quantitative and qualitative inquiry), and in its explicit recognition of the use of the self as the primary research instrument for documenting and interpreting the perspectives and experiences of the people and the cultures being studied. (1997, p. 13–14)

Ironically, there is some missing context here. One of the books that Lawrence-Lightfoot wrote after *The Good High School* was a personal and moving portrait of her mother, Dr. Margaret Morgan Lawrence (*Balm in Gilead: The Journey of a Healer*, 1988), a pioneering African-American child therapist whose life, despite her professional success, was deeply marked by racism. As one reviewer observed of the book:

> Her portrait is useful for filling in important figures missing from past historical accounts. These were Black people who, through unyielding effort and good luck, were able to prosper in Vicksburg and other places. Lightfoot has rescued them from what E. P. Thomson has called the enormous condescension of history. Knowledge of this group of Americans is invaluable for understanding the different ways that Blacks have made homes for themselves between the horns of the American dilemma: the contradiction between racism and the American

> creed. Lightfoot's figures wielded important influence in the Black community. They created institutions and built bridges that reached out to the larger world. (Featherstone, 1989, p. 370)

Evidently portraiture's "search for goodness" and interest in boundary-crossing, among other features, resonate with this larger social context of the author's own family history. Portraiture, we might say, is implicitly an *anti-racist* methodology. Indeed, in a special issue of *Qualitative Inquiry* in 2005, Lawrence-Lightfoot singled out the work of the iconic African-American intellectual W.E.B. Du Bois as her greatest source of inspiration: "He invented a way of being, a point of view, a style of work that quite naturally, dynamically, and organically integrated science, art, history, and activism" (2005, p. 10). Other contributors to the same issue emphasized the congruence of portraiture and critical race theory (Chapman, 2005) and the affinities between portraiture and jazz (Dixson, 2005), while still others offered portraits of African-American, Arab-American and White American educators illustrating "the ways in which race continues to matter in both positive and not-so-positive ways" (Dixson, Chapman, & Hill, 2005, p. 22).

Portraiture, in short, is very much at home in the ethnicized, racialized, gendered borderlands where resistance flourishes. Lawrence-Lightfoot's style of resistance is a subtle and graceful one; her emphasis on *seducing* the reader into greater critical awareness is characteristic. But one cannot fully make sense of the portraiture tradition without acknowledging its connections with a long emancipatory tradition in African-American culture, or with its fundamental orientation towards social justice and a compassionate feminist axiology.

This orientation was shared by Lawrence-Lightfoot's co-author in *The Art and Science of Portraiture*, Jessica Hoffman Davis; however, as an academic psychologist with a deep interest in the processes of art-making, she brought a different background and sensibility to the work. Alongside her colleague's emphasis on "the aesthetic principles of composition and form, rhythm, sequence, and metaphor" (1997, p. 12), Davis contributed insights into art as practice, and into portraiture as a kind of *thinking*: "In portraiture, as in any work of art, the medium is an agent of discovery" (1997, p. 36). Her inclusion in the book of "gesture drawings" by children and adults – quick, impressionistic "portraits" of emotions such as Angry, Sad and Happy – helped draw out the importance of "creativity and improvisation" in portraiture (1997, p. xvii) and aligned the methodology more explicitly with arts-based traditions in educational research.

Implicit in Davis's approach, though not foregrounded in the way it might be today, is the way that shared practices take on "a life of their own" – that is, both the artistic practice itself, and the objects it creates, become actors in the development of shared understandings among the participants. As will be seen, this is a salient feature of the methodology when it comes to ecoportraiture work. Davis was especially taken with the scope it offers for including a wide range of subjectivities; as she expressed it, portraiture could "be altered and recreated to respond to and embrace the differences among a group of researchers" (1997, p. xvii) – that is, to function as a methodology of collective interpretation and representation that avoids singularity of voice and gaze.

For her part, Lawrence-Lightfoot was explicit about the central role of relationships in portraiture:

> They are negotiated and renegotiated, week by week, day by day, even minute by minute as the portraitist and the actors navigate lines of intimacy, trust, reciprocity, and boundary setting, and as they work to develop a level of comfort, balance, honesty, and authenticity in their communications with one another. It is in the building of relationships that the portraitist experiences most pointedly the complex fusion of conceptual, methodological, emotional, and ethical challenges. (1997, p. 135)

Davis, however, pushed this conception farther, by describing the ways her group of portraitists interacted with five "exemplary" community arts centers, outlining "a process of documentation and evaluation that is explicitly activist and interventionist" (1997, p. 16). She encouraged portraitists to clarify their preoccupations and presumptions before entering the field, as a means towards cultivating "greater openness of mind" and the ability to receive participants' gifts of time, energy, access and trust humbly and attentively. They "must be diligent in their attempt to name convergence with the language of insiders" – that is, to align emerging themes with the authentic language and self-understandings of the actors, checking with the latter at regular intervals for confirmation and clarification.

At times, it may seem that the gaze of the portraitist is being privileged. In that sense the original metaphor contains an unhelpful bias, at least as we normally think of portrait painting, with the sitter placed in an essentially passive role. But a close reading of portraiture as a methodology shows that the gaze operates in both directions. The portraitist sees, but is also seen, and the way she is seen in the context of her relationship with the actors carries with it an ethical demand. The portrait must convey this *knowing gaze* of the actors: a gaze that is not content with knowing themselves and their own context, perhaps with greater clarity and insight than they did before, but that demands recognition from the viewer/reader just as it previously demanded such recognition of the portraitist.

The Art and Science of Portraiture was written at a time when qualitative research methodologies were just starting to establish themselves as an intellectually rigorous and respectable movement within the social sciences. Denzin and Lincoln's groundbreaking *Handbook of Qualitative Research* (1994) had appeared only three years earlier; Denzin's *Interpretive Ethnography* (1997) was published in the same year. The intervening two decades have been an extraordinarily fruitful time for the field; the Fifth Edition of the *Handbook* (Denzin & Lincoln, 2018) runs to nearly 1000 pages and, according to the publisher, includes "19 completely new chapter topics, including indigenous methodologies, methodologies in an age of new technologies, queer/quare theory, ethnodrama, data and its problematics, triangulation, collaborative inquiry, digital ethnography, the global audit culture, and much more" (SAGE Publishing, 2018). Seen in this context, portraiture no longer appears especially radical in its original formulation. Many of the themes taken up by Lawrence-Lightfoot and Davis are shared with other qualitative methodologies, while their emphasis on aesthetic unity and authenticity sits uneasily in an intellectual era of hybridity and contestation. The notion of "bridging art and science," in particular, can be read as an attempt to gloss over deep epistemological divides which today's researchers are more likely to tackle head on.

Yet the text remains generative. Some reasons for that are suggested in what has already been said. But there are more dimensions to explore.

Portraiture as Hermeneutics

Along with Du Bois, the introduction to *The Art and Science of Portraiture* acknowledges a debt to William James and John Dewey, two of the central figures of American Pragmatism. The allusions to James and Dewey are directly inspired by their rejection of formal abstraction in favor of "ways of envisioning and recording… experience that would not distort its texture and richness" (1997, p. 6). But this emphasis on learning through reflection on direct encounters with the world is not, of course, unique to Pragmatism. Dewey's exact contemporary Edmund Husserl sought something similar through what he termed "phenomenology," the study of how the world appears to us through experience. There are important differences between pragmatism and phenomenology, notably the latter's far more skeptical take on the truth claims of the natural sciences. Still, they share an appreciation of first-person experience as a way of knowing that, while partial, is not illusory, but rather fundamental to how the world actually is (Hills, 2017).

This phenomenological bent manifests itself in portraiture in a number of ways, perhaps most clearly in its emphasis on capturing the "essence" of the subject (Lawrence-Lightfoot & Davis, 1997, p. 4). "I felt the echoes of being on the other side of the artist's palette," Lawrence-Lightfoot writes; "I wanted the subjects to feel *seen* as I has felt seen—fully attended to, recognized, appreciated, respected, scrutinized" (pp. 4–5). As in phenomenology, the researcher understands such an essence as graspable in no other way than through the observer's subjectivity; the task is not to set this subjectivity aside, but to embrace what it can tell us, when placed "in close communion with rigorous and systematic attention to the details of social reality and human experience" (p. 8). The personhood and the voice of the researcher become self-evidently part of the research process, and skills in deliberate, contextual self-reflection become important for researchers to acquire.

Despite what was said earlier about the two-way gaze involved in portraiture, there remains what Chelstrom (2013) calls an "essential asymmetry" between the portraitist's experience and that of the participants she witnesses: the portraitist knows her own experience from the inside, but that of her subjects only through a kind of empathetic, imaginative resonance. Husserl himself emphasized that such experience of the alterity of others is necessarily "incomplete and non-totalizing" (Chelstrom, 2013, p. 16), laying the foundation for subsequent phenomenological explorations of how we subjectively encounter and relate to the Other (a vast and ramifying literature). More germane to portraiture, however, is the insight that each of us brings a vast range of experiential knowledge to bear in our efforts to interpret these encounters and relations; understanding always emerges within a context. Lawrence-Lightfoot spells out the implications:

> The portraitist… believes that human experience has meaning in a particular social, cultural, and historical context—a context where relationships are real, where the actors are familiar with the setting, where activity has purpose, where nothing is contrived (except for the somewhat intrusive presence of the researcher). … The researcher is the stranger, the one who must experience the newness, the awkwardness, the tentativeness that comes with approaching something unfamiliar, and must use the actors in the setting as guides, as authorities, as knowledge bearers. (1997, p. 43)

The portraitist, Lawrence-Lightfoot says, asks both "What is the meaning of this action, gesture, or communication to the actors in this setting?" and "What is the meaning of this to me?" (1997, p. 91). The potential gap between these perspectives points to how meaning in portraiture emerges *intersubjectively*: not simply in relation to another person (say, the subject of an interview), but in relation to a complex context in which both portraitist and portraitee are embedded, but differently situated. It is this reliance of understanding on our embeddeness in the life-world, a relationality-in-context that is *always already* in existence prior to any encounter with the Other, that marks a shift in emphasis between phenomenology and hermeneutics. Although references to the latter tradition are absent from *The Art and Science of Portraiture*, it may actually be the philosophical framework that is most closely attuned to portraiture's concerns and methods.

Hermeneutics, as Canadian curriculum scholar David G. Smith tells us (1991), originated in debates over the interpretation of sacred texts; over the course of several centuries, it became clear that the interpretive relationship between part and whole, text and context, applies not only to written texts or even to language itself but to all questions of human meaning. "One of the most important contributions hermeneutics makes to all contemporary social theory and practice," he suggests, "is in showing the way in which the meaning of anything is arrived at referentially and relationally rather than (for want of a better word) absolutely" (1991, p. 197). This is the reason why terms such as "conversation" and "tradition" crop up frequently in hermeneutic texts: living dialogs of this kind are exactly where insights and guidance are to be found. Indeed, modern hermeneutics has developed in a kind of dialog with phenomenology, as Smith observes: "From Husserl on, words like 'understanding,' 'interpretation,' and 'meaningfulness' are rooted, hermeneutically speaking, in a sense of the dialogical, intersubjective and conversational nature of human experience" (1991, p. 192).

The hermeneutic nature of portraiture is apparent in the emphasis placed by Lawrence-Lightfoot and Davis (1997) on context (see Chapter 3), "voice as the research instrument" (see Chapter 4), relationship (see Chapter 5), and on the emergence of meaning from "a dynamic process of receptivity, negotiation and accommodation" (1997, p. 186). Even the emphasis on aesthetics, foregrounded in Chapters 2 and 7, finds deep resonance in hermeneutic theory, especially in the writings of Hans-Georg Gadamer (e.g. 1986; 2004). The ability of a work of art to strike us, Gadamer says, is a measure of its truthfulness – its ability to disclose something about the world of which we are part while also shaping it. Thus the labor of the portraitist to "construct a portrait—an aesthetic whole—that resounds with authenticity" (1997, p. 247) is a search for hermeneutic truth of this kind.

Revealed as a form of hermeneutic scholarship, portraiture can also be more clearly seen as a kind of educational endeavor – for readers, participants, and researchers alike. As another Canadian scholar, David Jardine, has pointed out:

> It is not simply that hermeneutics is particularly well suited to education as an object of scholarly investigation. More than this, hermeneutics, as a way of conducting research, is educative in its intent. It wants to listen, to affect and to invite, not merely inform. ... Our choice of research methodologies is, of necessity, a choice regarding how understanding happens and what sort of voice is given to the "object" of such research. Therefore, our choice of research methodologies is a deeply pedagogical choice *at the outset*, before any research project has begun. (Jardine, 2015, pp. 269–270)

Is portraiture pedagogical in its intent? We can certainly interpret Lawrence-Lightfoot as saying so, not only in her 1997 book but in the essay that introduces the special issue of *Qualitative Inquiry* mentioned earlier. Here she writes about " a methodology... that seeks to inform and inspire and join the endeavors of documentation, interpretation, and intervention" (2005, p. 7); she suggests that portraitists "engage in acts (implicit and explicit) of social transformation [and] create opportunities for dialogue... This is provocative work that can disturb the natural rhythms of social reality and encounter; this is exciting work that can instigate positive and productive change" (2005, p. 12). Clearly, this is an engaged pedagogy, one acutely conscious of its ethical and political commitments as they arise from the search for goodness in a given social (and ecological) setting; that is, also, a deeply *situated* pedagogy.

The "choice regarding how understanding happens" is, in portraiture, a choice to work towards that understanding dialogically, empathetically, and inclusively. In a passage which echoes Jardine's claim, Lawrence-Lightfoot notes:

> Many of us are wanting to expand our audiences and welcome more voices into the public dialogues about education and schooling. If we want to broaden the audience for our work, then we must begin to speak in a language that is understandable, not exclusive and esoteric . . . a language that encourages identification, provokes debate, and invites reflection and action. But it is not only the language and idiom of our texts that will change, it is also that in anticipating different consumers of our work, we will begin to conceive of our research (the questions and design) differently from the very beginning. (2005, p. 9)

There is a long tradition in Western thought of hostility to ambiguity, ambivalence, and undecidability. This of course is reproduced in many ways in the processes and contents of formal education, but also, frequently, in cultural assumptions about what it means to be educated – a sense that education should ease our difficulties of understanding, much as medicine should ameliorate the problems of our bodies. Hermeneutics, on the other hand, treats occasions when our understanding of things seems to reach its limits as opportunities to enlarge our vision. "As Paulo Freire put it, reality is always 'hinged.' Reality is always reality-for-us but it always opens out into a broader world which serves or can serve to enrich our understanding of who we are" (Smith, 1991, p. 197). Portraiture shares this understanding. "One of the most powerful characteristics of portraiture," Lawrence-Lightfoot and Davis tell us, "is its ability to embrace contradictions, its ability to document the beautiful/ugly experiences that are so much a part of the texture of human development and social relationships" (1997, p. 9). This means that the results can sometimes be shocking to the actors themselves: "The images are rarely the ones they are used to seeing; they are often not very appealing or flattering" (1997, p. 260). Nonetheless, a successful portrait allows the actors to sense "the resonance in the essence of their being that gets revealed"(1997, p. 260).

Seeing portraiture as a kind of hinge-work, located at the joint between participants' conscious, limited self-understanding and a background of continually unfolding and limitless depth and complexity, aligns it with the work of Hermes, the divine trickster who by "rearticulating" the cosmos "keeps its articulated shape lively" (Hyde, 2010, p. 264). This, then, offers an image for the kind of pedagogy embodied in portraiture. It is a pedagogy of ongoing personal and cultural renewal, a "convivial struggle" to expand our understanding of "the real events of our lives and their possibilities" (Jardine, 2012, p. 7); a pedaogy of "free spaces" that are

> ...near-miraculous, imaginative, innovative, funny, challenging, enlivening, sometimes even enraging: places, broad, spreading, populated with myriad voices, and multiple, sometimes contradictory, witnesses; topics in that old, archaic sense that is at the heart of hermeneutics. (2012, p. 19)

Portraiture Beyond the Human

The foregoing description of portraiture offers several clues as to why, even though its original formulation is so clearly centered on the voices and doings of humans, it is more open than many methodologies to listening to the more-than-human. For a start, and in keeping with the phenomenological tradition, it is deeply interested in alterity. As noted earlier, portraiture's connections with the African-American experience are at the root of its concern to engage with voices that historically have been ignored or treated as deficient or marginal. The "search for goodness" is the work of coming to understand how others live meaningful lives *on their terms.* Expanding this concern to include the flourishing of other species, and the reciprocal and interdependent flourishing that characterizes "the wild" (Bringhurst, 2008), arguably requires a conscious shift of attention, but no fundamental changes in values or purpose. The "intentionally generous and eclectic process" of portraiture is well aligned with the striving of some phenomenologists "to open up to the experience of nature, or [to allow] 'nature' to speak to us via a different and more genuine or authentic register" (Vakoch & Castrillón, 2014, p. 1); that is, to engage with

> the unalterable alterity that is constitutive of our experience of nature. There is in the smell, taste, and feel of the wild, an uncanny otherness. This indelible and radical alterity acts as the mainspring for much of the vitality and vibrancy that we experience when immersed in the wild. (2014, p. 3)

Lawrence-Lightfoot herself notes that "portraits are always framed by the ecological context"; she suggests that the reader "should feel as if he or she is *there:* seeing the colours of the autumn leaves feeling the temperature of the gentle breeze, and hearing the rustling branches" (1997, p. 44). But as Jardine (1998), writing from a hermeneutic perspective, reminds us, such attentiveness prompts and invites us to go further. In opening up to the world as it exists and makes itself felt "over and above our wanting and doing," we can be jolted out of our self-absorption:

> Once I deeply realize that all these things are not here for me, they begin, so to speak, to turn away; they no longer pay any special attention to me; they are not formed up and tarted up for my perusal. Suddenly, I am no longer the displaced, post-Copernican, questioning stranger, the colonizing interloper, the condescending intruder to whom attention must be paid and who can make demands without attention to where I actually am, without attention to what is *already at work without me.* Suddenly, I *belong* here. I *live* here.... I am no longer the judge posing questions and demanding silence. I can begin, as ecology suggests, to become deeply conversant with things, listening, asking, responding, inhaling, and exhaling. Such conversation is only possible insofar as it issues up out of a sense of the integrity of my interlocutor— the Earth in all its wondrous articulations. (1998, pp. 99–100)

This is, as we have seen, the stance of the portraitist: "one of acceptance and discernment, generosity and challenge" (1997, p.141). Within an imaginative frame she calls "empathetic regard," Lawrence-Lightfoot describes listening to stories and deepening relationships in the research setting, learning to recognize insistent themes that offer "coherence, purpose and definition" to people's journeys, tracing this "life litany" as "a source of growth in work and identity" (1997, p. 197). To expand this work to include our more-than-human kin seems, in many respects, but another expression of the inclusive, just, and humble spirit of portraiture. "The portraitist comes to the actor as a supplicant, earnest and beseeching," she writes (1997, p. 155); and one need not search far in Jardine's work to find similar images of hermeneutic encounters with the more-than-human. We are, he says, "indebted to all things without exception"; "true thinking simply heightens our sense of this original debt, this original gift, this original blessing":

> I am not an *exception* to the interweaving, Earthly indebtedness and interdependency, but an instance of it. Believing this requires a deep sort of 'letting go' that is not commonplace in Western aspirations. It entails a sort of simple, reverential understanding of things and our reliances on them. . . . But this also occasions a realization of the fact that "the self in its original countenance" (Nishitani, 1982, p. 91) is an Earthly self, full of the humus that interweaves with all things. (1998, p. 98)

Yet if portraiture is hospitable to such insights, it can also be transformed by them. For example, if the "identity, character and history" of the researcher play a critical role in portraiture (1997, p. 13), then what does it mean to rethink them in ecological terms? What does it mean to include more-than-humans as actors/participants/co-researchers rather than just backgrounded gentle breezes and green leaves? If the portraitist is intent on recording "the subtle details of human experience" (1997, p. 14), what does it mean to think about that experience as deeply entwined within our kinship with the more-than-human? Or to record what our more-than-human kin is experiencing, including, at times, *us*? Such questions suggest that ecoportraiture is not, or at least not necessarily, simply portraiture that attends more deliberately to "the ecological context." The radical inclusivity noted earlier, and the commitment to a deeply situated pedagogy attuned to transformation, suggests that the portraitist may arrive at a different understanding altogether, in which the voices of the more-than-human are brought to the fore and *their* "search for goodness" made central to the research. At the least, the quest to extend portraiture beyond merely human concerns invites a substantial shift in perspective, a different quality of gaze, an approach to relationship that is more Earthly and humble than *The Art and Science of Portraiture* might suggest.

In *Place, Being, Resonance* (2015), Michael Derby (now Michael Datura, as in this volume) asks, among other things: "What if. . . we approached ecological education as a practice of learning to fix our attention on the resonant structure of meaning — the *voice* — of the world?" (2015, p. 9). "Voice in portraiture," Lawrence-Lightfoot tells us, "encompasses. . . epistemology, ideology, and method" (1997, p. 87), and so Derby's remark affirms the degree to which eco-hermeneutics and ecoportraiture can themselves be pedagogical, teaching new ways of witnessing, interpreting, mulling over, reflecting on, listening to and dialoguing with the human and the more-than-human. In addition, Derby is emphatic that this entails the development of an eco-poetic sensibility, involving "attentiveness, precision, and a tenderness towards existence" (2015, p. 32, citing Fisher-Worth and Street, 2013, p. xxvii). "The portraitist's 'urge for poetry'" is noted in an aside by Davis in *The Art and Science of Portraiture* (1997, p. 37); Derby helps us see how this demand can also be a quite literal one – that eco-poetic practice can help "restore ... the wisdom of phenomenological description and imaginative encirclement when attuned with ecological form" (2015, p. 133). Lawrence-Lightfoot's earlier remark about changes in "the language and idiom of our texts" in light of our intended audiences here takes on new meaning: does ecoportraiture also entail a search for "wild language" (Bringhurst, 2008), language more deeply resonant with the more-than-human?

From Jardine, Smith and others we know that the topic of a hermeneutic exploration helps define *how* that exploration can proceed; we know, too, that the personal journey of the explorer, including their ethical and political orientations, is intrinsic to what that exploration will uncover. In so far as the topic of ecoportraiture is always, in some sense, the embeddedness of human beings and doings in a more-than-human world, the processes of ecoportraiture must be responsive to and guided by that world in significant ways. As Jardine suggests, this is a kind of "letting go"; but one that simultaneously implies, as Derby argues, "a learned and storied discipline, a kind of wisdom requiring not only the ability to see-as, but the will and the psychological strength to consider what a tree means 'over and above our wanting and doing'" (2015, p. 140). Derby suggests that this in itself is an ethical stance; while "prefigurative political gestures" are not ruled out, critique is important primarily as a means "to clear space for deep relational possibilities and let the ethic emerge from within and follow of itself" (2015, p. 47). Of course, this kind of deep ethical relationality with the more-than-human may come more easily to some portraitists than others, for reasons both personal and cultural. Uncovering and reflecting on the portraitist's history of ecological relationality, or the way such relationality unfolds and deepens through the portraiture process, may be important to the work of ecoportraiture, just as it is in hermeneutics:

> The conversational quality of hermeneutic truth points to the requirement that any study carried on in the name of hermeneutics should provide a report of the researcher's own transformations undergone in the process of the inquiry; a showing of the dialogical journey, we might call it. Underscored here is a profoundly ethical aspect to hermeneutic inquiry in a life-world sense; namely, a requirement that a researcher be prepared to deepen her or his own self-understanding in the course of the research. (1991, p. 198)

As researchers in many places are coming to realize, this ecological self-understanding leads inexorably to a deeper engagement and self-questioning around issues of Indigeneity and (de)colonization (e.g. Root, 2010; Scully, 2012; Zinga & Styres, 2013). Ecoportraiture in many parts of the world is inevitably situated on colonized land. That colonial history may make itself known through absence – for example, the absence of Indigenous languages, traditions, and teachings based in the land – and also through many forms of presence, including colonial narratives of settlement and relationships to land and Indigenous peoples (Bhandar, 2018; Mackey, 2016; Tuck & Yang, 2012). Even when ecoportraiture does not directly involve Indigenous participants or an Indigenous portraitist, its work at the "hinge" of consciousness calls portraitists to deepen their awareness of this history, its enduring consequences for both human and more-than-human subjects, and the intersections of their work with local, regional and global struggles for justice and Indigenous resurgence.

Blenkinsop et al. (2016) suggest that colonialism has acted to suppress the voices of the more-than-human, placing eco-educators (and by implication, eco-portraitists) in an "interstitial zone" where a primary ethical imperative is to "shut up and listen" to what the colonized have to say. To this, a decolonizing approach to ecoportraiture might add a second imperative: to listen to what Indigenous knowledge holders have to say *about how to listen*. Ahenakew (2017, p. 81) comments that many conversations with Indigenous elders and educators "center around one deceivingly simple question: How can we (re)learn to listen to and be taught by the land, in the context of ongoing efforts to objectify, to commodify, to silence, or to speak for the land?" And even this, he says, does not do justice to the ways in which the land teaches through *being*. For this, Indigenous ceremonies and practice traditions may be a better guide:

> We cannot read, write, think, philosophize, empirically analyze, interpret, or shout our way out of onto-epistemic hegemony. Reigniting our sense of connection as part of the living land requires an interruption of the satisfaction we have with the boxes that provide our certainties and securities. We need to be bored with separability in order to be open to disinvest in what offers us comfort. The costs of being expanded into the flesh of the land and losing the illusion of independence through ceremonies like the Sun Dance need to outweigh our fear of vulnerability in interdependence. My prayer is that the land itself can dream us into a different future. (2017, p. 89)

In the face of such teachings, ecoportraiture must humbly acknowledge the inadequacy of existing frameworks, including those reviewed here, for understanding what must be done. In the end, the work of the eco-portraitist involves a kind of groping forward, guided by the experiences and intuitions of the *ecological* self, "empty of self-existence" as Jardine puts it, but "interlaced with all things" (1998, p. 93). Approached in this way, ecoportraiture becomes both profoundly ecological and profoundly personal. In *The Art and Science of Portraiture*, Lawrence-Lightfoot offers a short extract from one of her own portraits, featuring a "tall, dark, senior psychiatrist" co-leading a seminar on cultural sensitivity for third-year residents. "My voice as autobiography threads through this piece," she observes, "revealed in the nuances of our interactions, in the intensity of my questions, and in our raucous laughter that barely masks our pain" (1997, pp. 97–98). Pain, laughter, intense questioning, and deep conversation: not a bad image for the work to be done, as we move towards

> acts of true teaching and learning... that lift and open the heart despite its sorrows... not in order to turn away to some vaguely stupid Ecological Romance or equally distracting Ecological Panic. No, this lift allows *these very sorrows* to be sung in words and harmonies with ravens and stones and trees, and not just suffered in silence and isolation. (Jardine, in Derby, 2015, p. xix)

Steps into Ecoportraiture

Each of the chapters in this book tells a story of a more-or-less independent research journey, in which ecoportraiture, understood in various ways, emerged as a deeply helpful methodology for addressing themes of learning-and-teaching-in-place. Each of the authors has their own take on what ecoportraiture is or might be, shaped by their own autobiography and inquiry; each offers different illustrations of what the process looks like, and what its rewards and challenges are. Such diversity seems intrinsic to portraiture, as it must be to any methodology that engages with the particularity of places and beings, including the being of the researcher herself. Some of what portraiture has to teach us involves lingering with this diversity, letting the different stories and perspectives resonate with one another, letting hidden meanings emerge.

This is, of course, itself a hermeneutic process. Every reader will be struck differently by these stories, will find ethical or experiential or aesthetic or affective resonance in different details, will take away different lessons. The aim of this chapter has been to point to some of the deeper themes or patterns that the stories hold in common, but they are by no means "the whole story". Indeed, one of the lessons of hermeneutics is that when one "treats a topic well," giving it time and space and the kind of humble attention characteristic of portraiture, one becomes aware of its richness and vastness and inexhaustibility. This is one of the deep rewards of the kind of work described here; we hope that, to some degree, the book as a whole offers a similar experience to the reader.

Here is a short prefiguration of the chapters to come.

Nora Timmerman (Chapter 2) wanted to understand how ecological educators "understand and negotiate the concepts of integrity and contradiction in their day-to-day lives." She ended up focusing on three such educators – Madhu Prakash, David Greenwood and Ray Barnhardt – whom she visited and interviewed in the places where they lived and worked. In the course of writing up their portraits, she sought guidance from Jessica Hoffman Davis, aligning her work fairly closely with the methods and purposes of *The Art and Science of Portraiture.* She calls the resulting work "eco-ish portraiture," in which photography, drawing, and gathering natural objects from each subject's home region served to affirm the importance of ecological context. However, she also offers thoughts into how her research might have changed had it been conceived as ecoportraiture.

Kelly Keena (Chapter 3) undertook a study of children's interactions with an outdoor natural space adjacent to an elementary school south of Denver, Colorado. She was drawn to portraiture because of its "search for goodness," having come to realize that she was less interested in solving problems than in finding "a deeper understanding of how children saw themselves as part of something bigger than themselves, and when, or if, they isolated themselves apart from their natural surroundings." Her background as a naturalist helped her collect a rich harvest of observations and stories, illuminating not only who the children were *in that setting*, but also developing a portrait of the Habitat itself, as felt and seen through their experiences.

Laura Piersol (Chapter 4) was a member of the research team associated with the Maple Ridge Environmental School, a project involving all three of this book's editors (see also Datura, this volume). She was curious about how teachers' and students' relationships with the more-than-human world were changing in the context of spending every school day out in the forest. She ended up focusing on portraits of three particular children and one teacher at the Maple Ridge school. While she began with conventional sit-down interviews, she soon found herself accompanying the children she was working with on many kinds of explorations. Drawings, photography, audio and video all proved fruitful ways of expanding the inquiry and gaining insights into the children's observations, thoughts and feelings. At the same time she sought ways to welcome and acknowledge the presence of the more-than-human in the processes of teaching and research.

Sylvie de Grandpré (Chapter 5), an educator in southeastern BC whose practice is deeply informed by her Québécoise identity, was drawn to ecoportraiture as a means of exploring how outdoor education on Ktunaxa territory could become more deeply aligned with Indigenous understandings of Land. Her chapter charts resonances between ecoportraiture and Indigenous research "to see if there are enough points of connection for mutual enrichment." Moving through the themes of context, voice/stories, relationships, and processes/relational accountability, she does in fact find many parallels in the underlying values and purposes of these different traditions; however, she sees decolonization as a dynamic process that calls for portraiture, too, to be alert and open to its own limitations. In that vein she offers a few suggestions for "subtle adjustments" to portraiture methodology to make it more effective as a decolonizing practice and mode of representation.

Michael Datura (Chapter 6) joined the Maple Ridge Environmental School project at the start of 2013. On one level, his chapter unpacks some of the dilemmas and tensions encountered by the research team in trying to chart and make sense of the ways in which learning was shaped in the forested parks and other natural settings that hosted the school. As an illustration, he offers a fresh and funny depiction of "a day in the life" of the Environmental School that he terms a "gesture drawing," an example of "eco-friendly portraiture" that he proceeds to methodically and methodologically dissect. This in turn leads to deeper ruminations on the values, purposes and methods of eco-portrature, encouraging researchers to push the bounds of their own understandings and practices.

Finally, the editors offer an Epilogue in which they reflect, in conversation, on the potential of ecoportraiture as a way to share stories of teaching and learning across ecological settings. Not all forms of research hold such promise as an aesthetically engaging and convivial endeavor. We wish everyone success who finds inspiration in these pages, and look forward to learning from you in turn.

References

Abram, D. (2012). *The spell of the sensuous: Perception and language in a more-than-human world*. New York, NY: Vintage.

Ahenakew, C. (2017). Mapping and complicating conversations about Indigenous education. *Diaspora, Indigenous, and Minority Education*, *11*(2), 80–91.

Anderson, S. K. (2017). *Bringing life to school: Place-based education across the curriculum*. Lanham, MY: Rowman & Littlefield.

Berry, W. (2012). *A continuous harmony: Essays cultural and agricultural.* Counterpoint.

Bhandar, B. (2018). *Colonial lives of property: Law, land, and racial regimes of ownership.* Durham, NC: Duke University Press.

Blenkinsop, S., Derby, M., Piersol, L., & Affifi, R. (2016). Shut-up and listen: Implications and possibilities of Albert Memmi's characteristics of colonization upon the "natural world". *Studies in Philosophy and Education, 36*(3), 348–365.

Bringhurst, R. (2008). Wild language. In R. Bringhurst (Ed.), *The tree of meaning: Language, mind and ecology* (pp. 257–276). Berkeley, CA: Counterpoint.

Chapman, T. K. (2005). Expressions of "voice" in portraiture. *Qualitative Inquiry, 11*(1), 27–51. doi: https://doi.org/10.1177/1077800404270840

Chelstrom, E. (2013). *Social phenomenology: Husserl, intersubjectivity, and collective intentionality.* Lanham, MA: Lexington Books.

Denzin, N. K. (1997). *Interpretive ethnography: Ethnographic practices for the 21st century.* Newbury Park, CA: SAGE.

Denzin, N., & Lincoln, Y. (Eds.). (2018). *Handbook of qualitative research* (5th ed.). Newbury Park, CA: SAGE.

Denzin, N., & Lincoln, Y. (Eds.). (1994). *Handbook of qualitative research.* Newbury Park, CA: SAGE.

Derby, M. (2015). *Place, being, resonance: Towards a critical ecohermeneutic approach to education.* New York, NY: Peter Lang.

Dillard, A. (2015). *Pilgrim at Tinker Creek.* Norwich, UK: Canterbury.

Dixson, A. D. (2005). Extending the metaphor: Notions of jazz in portraiture. *Qualitative Inquiry, 11*(1), 106–137. doi: https://doi.org/10.1177/1077800404270839

Dixson, A. D., Chapman, T. K., & Hill, D. A. (2005). Research as an aesthetic process: Extending the portraiture methodology. *Qualitative Inquiry, 11*(1), 16–26. doi: https://doi.org/10.1177/1077800404270836

Evernden, L. L. N. (1993). *The natural alien: Humankind and environment.* Toronto, ON: University of Toronto Press.

Fisher-Worth, A., & Street, L.-G. (Eds.). (2013). *The ecopoetry anthology.* San Antonio, TX: Trinity University Press.

Gadamer, H. (1986). *The relevance of the beautiful and other essays.* Trans. N. Walker; Ed. R. Bernasconi. Cambridge, UK: Cambridge University Press.

Gadamer, H. (2004). *Truth and method.* Trans. & Ed. J. Weinsheimer & D. G. Marshall (2nd rev. ed.). New York, NY: Continuum.

Garrard, G. (2012). *Ecocriticism* (2nd ed.). Abingdon, Oxon; New York, NY: Routledge.

Garrard, G. (Ed.). (2014). *The Oxford handbook of ecocriticism.* Oxford, UK: Oxford University Press.

Gough, A. (2013). Historical, contextual, and theoretical orientations that have shaped environmental education research. In R. B. Stevenson et al. (Eds.), *International handbook of research on environmental education* (pp. 9–12). New York, NY: Routledge.

Hills, J. L. (2013). Pragmatism and phenomenology: A reconciliation. *American Catholic Philosophical Quarterly, 87*(2), 311–320.

Hyde, L. (2010). *Trickster makes this world: Mischief, myth, and art.* Foreword by M. Chabon. New York, NY: Farrar, Straus and Giroux.

Jardine, D. W. (1998). *To dwell with a boundless heart: Essays in curriculum thoery, hermeneutics, and the ecological imagination.* New York, NY: Peter Lang.

Jardine, D. W. (2012). *Pedagogy left in peace. Cultivating free spaces in teaching and learning.* New York, NY: Continuum.

Jardine, D. W. (2015). On hermeneutics: "Over and above our wanting and doing". In J. Kincheloe & K. Tobin (Eds.), *Doing educational research* (2nd ed., pp. 235–254). Leiden, NE: Brill Sense.

Kimmerer, R. W. (2013). *Braiding sweetgrass: Indigenous wisdom, scientific knowledge and the teachings of plants.* Minneapolis, MN: Milkweed.

King, T. (2003). *The truth about stories: A native narrative.* Toronto, ON: House of Anansi.

Knight, S. (Ed.). (2013). *International perspectives on forest school: Natural spaces to play and learn.* Newbury Park, CA: SAGE.

Lawrence-Lightfoot, S. (2005). Reflections on portraiture: A dialogue between art and science. *Qualitative Inquiry, 11*(1), 3–15.

Lawrence-Lightfoot, S., & Davis, J. H. (1997). *The art and science of portraiture.* San Francisco, CA: Jossey-Bass.

Mackey, E. (2016). *Unsettled expectations: Uncertainty, land and settler decolonization.* Halifax, NS: Fernwood.

McCoy, K., Tuck, E., & McKenzie, M. (Eds.). (2017). *Land education: Rethinking pedagogies of place from indigenous, postcolonial, and decolonizing perspectives.* New York, NY: Routledge.

Nishitani, K. (1982). *Religion and nothingness.* Berkley, CA: University of California Press.

Oliver, M. (2013). *New and selected poems, volume one.* Cambridge, MA: Beacon Press.

Payne, D. G. (1996). *Voices in the wilderness: American nature writing and environmental politics.* Lebanon, NH: University Press of New England.

Reid, A. D., & Payne, P. G. (2013). Handbooks of environmental education research: For further reading and writing. In R. B. Stevenson, et al. (Eds.), *International handbook of research on environmental education* (pp. 529–541). New York, NY: Routledge.

Renshaw, P., & Tooth, R. (Eds.). (2017). *Diverse pedagogies of place: Educating students in and for local and global environments.* New York, NY: Routledge.

Rigby, K. (2014). Romanticism and ecocriticism. In G. Garrard (Ed.), *The Oxford handbook of ecocriticism* (pp. 60–79). Oxford, UK: Oxford University Press.

Root, E. (2010). This land Is our land? This land is your land: The decolonizing journeys of white outdoor environmental educators. *Canadian Journal of Environmental Education, 15,* 103–119.

SAGE Publishing. (2018). *The SAGE handbook of qualitative research, fifth edition: Key features.* Accessed at https://us.sagepub.com/en-us/nam/the-sage-handbook-of-qualitative-research/book242504#features

Scully, A. (2012). Decolonization, reinhabitation and reconciliation: Aboriginal and place-based education. *Canadian Journal of Environmental Education*, *17*, 148–158.

Selin, H. (Ed.). (2013). *Nature across cultures: Views of nature and the environment in non-western cultures.* Dordrecht, NE: Springer Science & Business Media.

Smith, D. G. (1991). Hermeneutic inquiry: The hermeneutic imagination and the pedagogic text. In E. Short (Ed.), *Forms of curriculum inquiry.* Albany, NY: SUNY Press.

Smith, G., & Sobel, D. (2010). *Place- and community-based education in schools.* New York, NY: Routledge.

Sobel, D. (Ed.). (2016). *Nature preschools and forest kindergartens: The handbook for outdoor learning.* St. Paul, MN: Redleaf.

Stevenson, R. B., Brody, M., Dillon, J., & Wals, A. E. (Eds.). (2013). *International handbook of research on environmental education.* New York, NY: Routledge.

Stevenson, R. B., Dillon, J., Wals, A. E., & Brody, M. (2013). The evolving characteristics of environmental education research. In R. B. Stevenson, et al. (Eds.), *International handbook of research on environmental education* (pp. 512–517). New York, NY: Routledge.

Stibbe, A. (2015). *Ecolinguistics: Language, ecology and the stories we live by.* New York, NY: Routledge.

Styres, S. D. (2017). *Pathways for remembering and recognizing indigenous thought in education: Philosophies of Iethi'nihsténha Ohwentsia'kékha (land).* Toronto, ON: University of Toronto Press.

Tuck, E., & Yang, K. W. (2012). Decolonization is not a metaphor. *Decolonization: Indigeneity, Education & Society*, 1(1), 1–40.

Vakoch, D. A, & Castrillón, F. (Eds.). (2014). *Ecopsychology, phenomenology, and the environment: The experience of nature.* New York, NY: Springer.

Wagoner, D. (1999). *Traveling light: Collected and new poems.* Champaign, IL: University of Illinois Press.

Waite, S. (Ed.). (2011). Children learning outside the classroom: From birth to eleven. Newbury Park, CA: SAGE.

Zapf, H. (2016). *Handbook of ecocriticism and cultural ecology.* Berlin, Germany: De Gruyter.

Zinga, D., & Styres, S. (2013). The community-first land-centred theoretical framework: Bringing a 'good mind' to Indigenous education research? *Canadian Journal of Education*, 36(2), 284–313.

2. *Relationship, Complexity, Artistry, and Co-Creation in Portraiture Research*

Nora Timmerman

Happy Valley

Figure 2.1. Happy Valley in autumn. Photograph by Nora Timmerman

Happy Valley is a name used by locals to refer to the State College region of inner Pennsylvania. Penn State, a land-grant university, has been the hub of this region since its establishment as an agricultural college in 1855.

Exhausted after a red-eye flight across the continent, I gaze out the window as our plane descends into the autumn-colored, rolling hills of Happy Valley, Pennsylvania. A brisk walk into, through, and out of the tiny airport, followed by a welcome inhale of balmy fall air, begins to waken my senses. Although I do not see her right away, my exhaustion is replaced with excitement and anticipation as I scan the sidewalks and parking lot for my host. A short two minutes of sitting in the sunny grass later, I catch sight of bold reds, oranges, and blacks lightly blowing in the breeze. Laughing with affection and joy before we even make eye contact, a vivacious woman—beautiful and radiant—walks toward me with arms extended.

Full Professor at the Pennsylvania State University, Madhu Suri Prakash would, over the course of the next week that I spend with her, describe herself in various ways: a philosopher of education, yes. But also: a daughter, a mother, a loner, a friend, a Luddite, and, simply, soil. I came to visit Madhu because of her unique perspective and voice within the field of eco/environmental education, her experience therein, and her animated interest in participating in this research.

Madhu had insisted on picking me up at the airport as if there would have been no other option. After our initial hug and "nice to finally meet you!" we walk toward her car as she asks how to best make me comfortable, rested, fed, and cared for. I offer mixed answers of both thanks and polite deferrals. Stopping briefly at her home, I discover that dinner is already prepared for the evening, she has purchased groceries for me for the week, and she makes me breakfast as I use her phone to call home. Madhu tells me that in her culture, guests are god; if you are not the best host possible, you are not showing respect to divinity. Guests need to know that they matter. Sipping warm tea, hunched over a bowl of freshly cut fruit and yogurt, I sit at her table wearing her red silk slippers. I am utterly wrapped in hospitality. Half an hour later, Madhu drops me off to rest with my groceries and a box of Bengal spice tea that she tossed on top as we headed out the door.

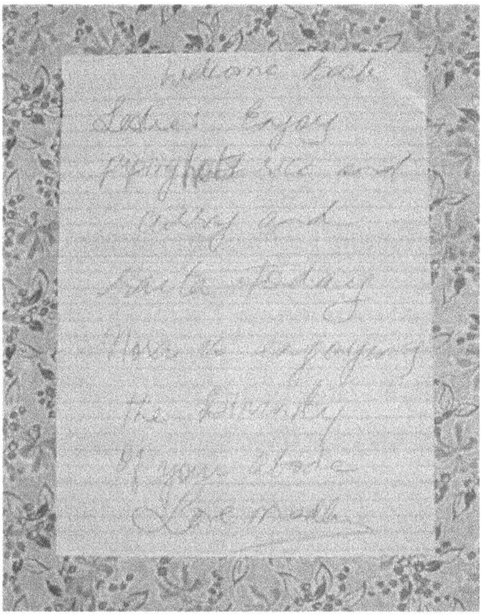

Figure 2.2. Joyful note. Photograph by Nora Timmerman

A note from Madhu to her good friend Leslie, with whom I stayed while I was in town: "Leslie: Welcome back! Enjoy piping hot rice and curry and raita today. Nora is enjoying the Divinity of your abode. Love, Madhu"

How I Came to Portraiture

It was the fall of 2010 when I visited Madhu. She was the second of three postsecondary educators that I visited for my doctoral research, a portraiture-based study (Timmerman, 2013). The short story above is an excerpt from one of the three portraits that I wrote for the dissertation, and I open this chapter with these words to give a sense of what portraiture looked like for me. I share excerpts from the other portraits throughout this chapter as I work to explain my understandings and use of portraiture and the possibility for *eco*-portraiture.

To understand portraiture, it may be most helpful to begin by explaining why it is used, what goals it sets out to meet. For my own research, I explored how educators understand and negotiate the concepts of integrity and contradiction in their day-to-day lives. I wanted to know how they felt, named, and engaged with the struggle to live according to their values in a world where the cultural norms and political-economic structures they live within are at-odds with those values. I felt strongly that the results of my research must not be prescriptive and could not be generalizable. I hoped that my research would offer inspiration and a sense of possibilities (for myself and others), but I also did not want to frame my research as an "answer" to the question of "how to live with integrity." The purpose of my research was thus to inquire into, critically explore, and share thoughtful possibilities for teaching and living with ecological integrity.

I aligned myself with scholars and activists who were (and are) striving for a less anthropocentric and more ecologically engaged, respectful, responsible, and reciprocal world, specifically those working toward these aims within postsecondary educational institutions. I wanted my research to offer a space for educators to reflect on their lives and practice, as Barone (2000) suggests, to conspire with participants' stories. He explains:

> Conspiracy can be a profoundly ethical and moral undertaking. "Conspiracy: combination or union of persons for a single purpose or end" (Webster's New Collegiate Dictionary). From the Latin, "con plus spiare, to breathe together," or better, from the Old French, "conspirer, a learned borrowing" (World Book Dictionary). A conspiracy, thus, is a conversation about the relationship between present and future worlds. The reader, a historically situated self, learns from the re-created Others in the text to see features of a social reality that may have gone previously unnoticed. . . . There is a "breathing together," a sharing of ideas and ideals for the purposes of an improved reality. (pp. 145–146)

I was also seeking out a research methodology that would allow me to convey complexity and ambiguity. I felt complexity and ambiguity myself, and knew that any realistic representation of integrity and contradiction would need to allow for the expression of them.

Portraiture was the right methodological approach to help me reach these goals. It is emotive, invites vulnerability, allows for complexity, and is (hopefully) enjoyable to read. I had not heard of portraiture before engaging in my doctoral studies, and it remains a relatively unknown approach to research. My dissertation co-advisor had worked with Jessica Hoffmann Davis, one of the co-authors of the "manual" on portraiture, for his own doctoral work, and he suggested that I look into it. Upon doing so, I saw the ways it fit with my overall research goals, but I also became viscerally excited by the possibility of using portraiture. It would allow me to meet and have fantastically rich conversations with widely respected educators. Portraiture was thus both a theoretically coherent and personally compelling methodology for my work.

In this chapter I share descriptions of portraiture itself, examples of how I engaged with this methodology, and an exploration of the idea of "eco-portraiture." Because portraiture is all about stories, I use story to explain much of my thinking and also interlace the chapter with excerpts from portraits that I have written.

Portraiture is a form of inquiry that asks researchers to form deep relationships with participants, explore participants' experiences and ideas about particular phenomena, and write those experiences up into literary "portraits" of participants. Sara Lawrence-Lightfoot coined the term "portraiture" in the 1990s. Lawrence-Lightfoot and Hoffmann Davis (1997) explain that, "Portraits are designed to capture the richness, complexity, and dimensionality of human experience in social and cultural context, conveying the perspectives of the people who are negotiating those experiences" (p. 3). Thus, while portraits themselves are often of individual people or of individual learning sites, portraitists dive into specific experiences with the goal of tapping into something that resonates much more generally. Lawrence-Lightfoot and Hoffmann Davis demonstrate how storytellers and social scientists alike take this same approach:

> Eudora Welty (1983) offers a wonderful insight gained from her experience as a storyteller. She says forcefully: "What discoveries I have made in the process of writing stories, all begin with the particular, never the general." Clifford Geertz (1973) puts it another way when he refers to the paradoxical experience of theory development, the emergence of concepts from the gathering of specific detail. Geertz (1973) says, "Small facts are the grist for the social theory mill" (p. 23). The scientist and the artist are both claiming that *in the particular resides the general.* (p. 14)

Lawrence-Lightfoot and Hoffmann Davis (1997) look specifically to storytellers and ethnographers because portraiture itself is built upon a combination of the more recognized methods of ethnography and narrative inquiry. Portraiture is ethnographic through its immersive, descriptive (versus experimental) fieldwork, its interest in a wide variety of meaningful aspects of participants' lives (versus only pre-determined foci), and its categorical, thematic forms of analysis (Goetz & LeCompte, 1984). It is similar to narrative inquiry in its emphasis on aesthetic, story-based interpretation and representation (Clandinin, Pushor, & Orr, 2007; Hart, 2002; Richardson, 1997).

Lawrence-Lightfoot and Hoffmann Davis (1997) see the combination of ethnography and narrative inquiry as a way to bridge "the realms of science and art, merging the systematic and careful description of good ethnography with the evocative resonance of fine literature" (p. 6). They describe portraiture as an approach that brings together the "two worlds" of art and science, "allowing for both contrast and coexistence, counterpoint and harmony... allowing [us] to see clearly the art in the development of science and the science in the making of art" (p. 3).

This blending is appealing to my own sensibility as a researcher who has always been interested in arts-based forms of research. The ideal of creating a "systematic and careful description" with the "evocative resonance of fine literature" aligned with my goals of creating research that was inspiring for others to read. Yet, alongside this appreciation, I questioned the way in which Lawrence-Lightfoot and Hoffmann Davis describe art and science as two separate "worlds," erring on the side of not challenging the dominant associations of science with objectivity and art with subjectivity. In my work, I drew upon others to suggest that both art and science are subjective and aesthetic in their own right (Clough, 2002; Eisner, 1997; Richardson, 1997). That is, the differences between art and science are less about inherent qualities, but are more about different aims. As Barone (2001) explains, while science is interested in "reducing uncertainty," art is more interested in "uncovering and expressing alternate (sometimes even conflicting) interpretations of the phenomena under scrutiny" (p. 24).

Portraiture does not seek validity in the form of objectivity or generalizability. Instead, one of the key goals of portraiture, and what Lawrence-Lightfoot and Hoffmann Davis (1997) say is an indication of "valid" portraiture, is a portrait that expresses the "authenticity" and "essence" of participants. Even though they fully acknowledge the overwhelming influence of the researcher in shaping the purpose, questions, and representation of the project, their goal is for participants to have a "Yes, that's me!" response when they read the final portraits (p. 247). This goal has been the subject of some critique, as it could suggest some kind of finality to a person, a capital-T "truth," or the sense that people do not change with time and experience (English, 2000; McKenzie, 2006). In my own experiences with portraiture, I often wondered – even if I were to accept the notion that each person has an essence – whether I, as a researcher, would even have the capacity to sense, identify, translate, and represent that essence. Thus, in response to these concerns over terms, I used Lawrence-Lightfoot and Hoffman Davis' (1997) discussion of "authenticity" and "essence" as a cautionary guideline for researchers. In seeking "authenticity" and "essence," they rightly warn of the potential to create simple (versus complex) portraits and/or portraits that represent the researcher more than the participant.

In my work, I attempted to take heed of their warnings while also avoiding essentializing. I did this through two equally important processes: 1) establishing a "multi-layered resonance," and 2) establishing practices for the inclusion of ambiguity and multiple "truths" about participants. A multi-layered resonance evaluates whether portraits create representations of participants' lives that resonate with participants themselves, with the researcher(s), and with readers (Clandinin, Pushor, & Orr, 2007; Clough, 2002; Hart, 2002; Lawrence-Lightfoot & Hoffmann Davis, 1997). To find this resonance, I created many versions of each portrait and shared selected versions of those with my doctoral co-advisors. These multiple versions and readings gave me space from my writing to evaluate it for "resonance," something that can be difficult when one is close and attached to one particular version of a story. Each participant also received drafts of their respective portraits for review. Before this, however, I asked Jessica Hoffmann Davis for advice about how best to go about sharing draft portraits with my participants. She suggested that I give participants two opportunities for feedback: one once I had a solid first draft that myself and my committee felt good about. This was a chance for participants to give substantive, major feedback. It was difficult. I broke down in tears on a phone call with Madhu discussing how I had interpreted something wrong. Yet, even this difficulty was one side of a coin of our relationship; it was hard because she trusted me and because I respected her. After this substantive round of feedback, the second was editorial, fact-checking types of edits toward the end of the writing. Jessica Hoffmann Davis suggested limiting participant edits to these two rounds because it would be easy to otherwise end up in a cycle of revisions that attempted somehow to erase the portraitist's presence. She argued, and I agreed, that the portraits are ultimately created by researchers, and while they ought to resonate with participants, they are not self-portraits.

Alongside the search for "multi-layered resonance," I also included stories from multiple voices and standpoints within portraits, trying to illustrate how the portrait was formed through a complex intertwining of voices and stories. In the following sections, I describe this process further, exploring what it means for portraiture to be a "search for goodness" and the implications of its foundation in storytelling.

A Faustian Bargain

A short three-minute walk up the country road in front of David Greenwood's house brings us to Greenwood Cemetery. Continuing only one minute more would take us to the wheat fields. Before we get to the wheat and turn onto the winding road that skirts the Palouse River, David walks up to a gravestone. It is simple: gray, rectangular, horizontally set, typical at first sight. As I read the epitaph for the second time, however, I puzzle over its meaning. "Victim of corporation greed," it says. Recalling the details from a local, historical publication, David shares the story.

Figure 2.3. Victim of corporation greed. Photograph by Nora Timmerman

In the early 1900s, Palouse was the hub of logging, mining, and railroad industries working as fast as they could to make as much money as they could from the land. Yet, these industries were starting to dry up and the economy was shifting to agriculture. As businesses fought to stay alive, they scraped corners, overworked employees, and became lax on safety regulations. The man whose grave we stood upon worked for Northern Pacific Railroad. One day, while filling an oil can near the repair tracks, a "door from a ballast car swung out and caught young Brown, crushing him against the building where he was at work, breaking the pelvis bone and inflicting internal injuries" (Kiessling, 1999, p. 23). Especially tragic as he had only a few days prior decided to "take up some less hazardous line of work" (p. 24) in light of his upcoming marriage, Bert Leon Brown died at age 24 of internal injuries, "victim of corporation greed." His parents, who had just barely made it to the hospital to spend his last 40 minutes of life with him, were irate with the injustice of their son's death and chose these words to adorn his gravestone lest the people of the Palouse forget.

As David finishes telling Bert's story, he tells me that working in the university sometimes feels like working in a corporation. Even alongside the many privileges and freedoms that come with being a professor of education, one can also sometimes feel a "victim of corporation greed." Yet, this feeling is a microcosm of the larger context of empire, the primacy of corporate power over people and land. All life (humans, forests, birds, frogs, and rivers, the Earth itself) is a "victim of corporation greed." Privileging output over worker safety, Bert's death is symptomatic of a culture of instrumentalism, a culture of what, in reference to Hardt and Negri's (2000) work, David calls "empire." He writes,

> Like globalization, empire describes the political economy of the planet: the new imperialism, colonization, development, free trade. Empire—a system of domination and resistance, a bio-political power that is exercised, internalized, and shaped by networks of human cultures worldwide. A system of authority and control enacted by all of us, motivated by habit, addiction, desire, necessity, dreams of a better life, fantasies of endless economic growth... the subject of empire is the commodity; the object is the consumer. All of us are its soldiers. (2010b, p. 14)

In the wake of empire are countless victims, some worse off than others, and some more complicit than others. David offers a quote from Wendell Berry (2012) in agreement that empire is "indifferent... It [does] not *intend* to victimize its victims. It simply follow[s] its single purpose of the highest possible profit, and ignore[s] the 'side effects' " (para. 21).

Talking and walking together this fall morning, I am struck by the depth and detail of David's historical knowledge of this place. This knowledge is what brought us to Bert's gravestone. And, the story of Bert's unfortunate death provided a frame through which we explored David's perspectives on socio-ecological "victimization," and "corporate greed," as well as his sense of personal complicity and resistance therein. Fully acknowledging educational institutions' contribution to socio-ecological injustices, David has understandably mixed feelings about his work. On one hand, he says, it's the greatest job ever, but on the other, it's the Faustian bargain, it's like you're selling your soul at the crossroads.

Seeking a Complex and Ambiguous "goodness"

One of the foundational assumptions of portraiture is that it is a "search for goodness." Most of my colleagues raise their eyebrows when they hear this phrase; it requires some explaining. Seeking out "goodness" means that the research is used to better understand what "good" means in different contexts. That is, portraiture seeks context-specific and fluid definitions of "goodness;" it is intended to open up a discussion on what "good" means (relative to the research questions) with participants:

> [Portraitists] are concerned with documenting how the subjects or actors in the setting define goodness. The portraitist does not impose her definition of "good" on the inquiry, or assume that there is a singular definition shared by all... Rather the portraitist believes that there are myriad ways in which

> goodness can be expressed and tries to identify and document the actors' perspectives. (Lawrence-Lightfoot & Hoffmann Davis, 1997, p. 9)

Thus, seeking out goodness does not mean that researchers "idealize… human experience [or] focus only on good things, look only on the bright side, or give a positive spin to every experience" (p. 141). By seeking out goodness, portraiture actually resists universalized, static definitions of what is "good" and turns toward relational, contextual understandings of goodness, making them the very focus of the research itself.

On one hand, my research aligned well with Lawrence-Lightfoot and Hoffmann Davis' descriptions of a "search for goodness." Like the authors describe, I simultaneously wanted to learn from participants' experiences and understandings, *and* interrogate those understandings *with* them. I approached participants as leaders, as whole human beings with faults, but as exemplars, mentors, people from whom I had something to learn and with whom I could deeply engage and interrogate ideas.

On the other hand, my research question and purpose pushed me toward a search for goodness that was more complex and ambiguous because I also sought out the – for lack of a better phrase – *lack* of goodness. I wanted to learn how other people experience and explain both "integrity" *and* "contradiction" in their day-to-day lives. Integrity is often considered "good" and contradiction "bad." Alternatively, I understand them dialectically, connected, and always a part of the same coin. Thus, my approach to portraiture's seeking goodness saw me seeking vulnerability, paradox, complicity, in the same breath as I sought honesty, alignment, commitment.

To "seek goodness" in ways that reveal not only strengths, but weaknesses, not only confidence, but vulnerability, it was essential to establish trusting relationships with participants. Portraits are often not anonymous, and the data collection, analysis, and representation process is a long one. It is one thing to explore the moral ambiguity of our experiences as anonymous people in an anonymous setting in a once-off meeting, but when complex aspects of our lives are explored at length and made public, trusting, ethical relationships are paramount. Ethics of care and responsibility thus shape the relationship between portraitists and participants themselves, and between portraitists and the final research portraits.

In my experience, the creation and renewal of trusting relationships was one of the most powerful and joyous characteristics of portraiture. When I walked to that gravestone with David and started asking him about his sense of his own complicity, I was incredibly nervous. We had had at least a couple phone conversations and a few email exchanges prior to that, but we hadn't yet laughed together at length, or shared a meal, we had just barely begun the process of revealing parts of our selves that deepen a relationship. When research depends upon a relationship, the shape of that research changes. I was not able to solely ask David my questions and leave. My use of portraiture called for reciprocity: if I ask you to share something vulnerable, then I need to be willing to share too. We did not engage in a kind of tit-for-tat, back-and-forth, but for all of my research participants, much of the time we spent together, we spent talking and getting to know each other. It felt only fair that they got to ask me about my family, my experience, and my thoughts too. Now, seven years after I visited my participants for the first time, I am no longer nervous. All that sharing has meant that we easily laugh together, cry together, and certainly cherish the opportunity to eat together.

The trusting relationships my participants and I built were thus important for diving into complex, vulnerable topics, but they were also necessary for creating the portraits (Lawrence-Lightfoot & Hoffmann Davis, 1997). Specifically, attending to these relationships helps researchers as they engage in the complex work of identifying voice, storylines, and themes. The practice of nurturing a relationship allows researchers to acknowledge the co-creation of knowledge between participants, researchers, and their (human and natural) communities. That is, through the creation and renewal of relationships between portraitists and participants, researchers are compelled to analyze and represent portraits as forms of shared/co-created knowledge.

Whereas many research approaches might call for separation between researcher and participant, identifying objectivity as the primary indicator of validity, portraiture instead fosters values of interconnection, responsibility, and care. One benefit of nurturing these values is that the resulting relationships open up the possibility for a fuller expression of complexity and ambiguity in the research.

An Appropriate Case of Laryngitis

On my last night in Alaska, we had dinner at the Fairbanks Pump House restaurant—myself, Ray, his wife Carol, and two visiting professors from New Zealand, one of which was visiting Fairbanks to give a presentation at the Alaska Federation of Natives conference titled, "The Role of Education in the Empowerment of a People." In that presentation, he talked largely about the contribution of education to the "cultural and economic enhancement" of Indigenous communities. During dinner, as he discussed the sacredness of the land to Maori people, I asked him about his earlier presentation, "Don't you find that with economic development, there's harm to the land?" Right away he answered, "Yes, that's one of our dilemmas." And, after explaining

Figure 2.4. Family in Alaska. Photograph by Nora Timmerman

"The First Unknown Family," statue of a man, woman, child, baby and dog at the centre of Fairbanks' Golden Heart Park on the banks of the Chena River.

various considerations and ways in which this dilemma manifests itself, he concluded, "We're looking for a middle ground: economic development through cultural means so that we don't degrade the land; there's a lot of challenges to that, but that's the goal."

That evening over dinner, I had many questions about these ideas and goals. Mostly, I lean toward visions of the future that seek non-capitalist futures, that critique the very notion of "economic development," as I see it being the root of the same land degradation the speaker wants to prevent. Yet, I do not know this person well. He is an Indigenous Elder, me a young, white grad student. Moreover, I have not extensively studied or immersed myself within the struggles for decolonization and Indigenous autonomy. Perhaps if we were in relationship, we could have this dialog, and perhaps we would find that we do not disagree at all. As it is, I listen to understand and remain silent. Perhaps it is somewhat fitting, then, that over the course of my Alaska trip, I have laryngitis. I stay on the outskirts of gatherings to avoid passing on germs, and mostly observe people and places. When I do talk, it is in a physically timid voice, hoarse from coughing, yet poignantly indicative of the uncertainty I feel within.

Ray Barnhardt is also white, a settler in Alaska, born from second-generation immigrant parents. But, unlike me, he has made it his life's work to know about and work toward Indigenous autonomy in education. He has lived in Fairbanks, Alaska for more than 40 years. Dozens of initiatives and projects list his name as collaborator with First Alaskan Elders, researchers, administrators, and teachers. In fact, even most of Ray's nuclear family has significant ties to First Alaskan education. Many years ago, when they first moved to Alaska, Ray traveled with Carol and their two-year-old son to rural villages as a field researcher for a study on "American Indian education." Navigating the troubled waters of non-Indigenous research within rural Indigenous communities, having his family present—particularly his young son—"broke the ice" for Ray, opening the doors for trust and connection. Through family, Ray was able to make his entrance into First Alaskan culture and forge relationships that continue to this day. At the time of my visit with Ray, Carol is the Chair of elementary teacher education for the University of Alaska Fairbanks (UAF). Their daughter, Amy, is an elementary teacher and adjunct professor of education at UAF. And Ray's grandchildren attend the place-based Watershed Charter School that bases its curriculum on the Alaska Standards for Culturally Responsive Schools. First Alaskan education is something Ray's whole family knows about in intimate ways, even if they themselves are not Indigenous.

And yet, my visit to Alaska reveals traces of uncertainty in Ray as well. It is difficult for me to get Ray to talk about himself. When I ask him about his own ideas or the challenges he faces personally, he tends to turn the conversation toward concrete examples of projects he has worked on, or if he does talk about ideas, they are often the ideas of others. This tendency seems partly indicative of Ray's immersion in Indigenous communities where relational, place-based, and collaborative understandings of knowledge are prevalent. Still, I think there is something else going on. Ray's disinclination to talk about himself and his ideas feels like a distant cousin to my own actual and metaphorical laryngitis. As a non-Indigenous person working in and with Indigenous communities, Ray must continually re-negotiate his own role according to the unfolding needs of Indigenous educators in their work toward autonomy. Ray's work, then, is largely about positioning. What I initially perceive as silence or uncertainty is at least partially attributable to a lived positionality of respect, deference, and humility.

Co-creating Stories with Artistry

Storytelling is foundational to the process and product of portraiture; story is used to both seek out and share knowledge. In general, listening for and telling stories is simply what we do, as human beings: "Whoever you are, wherever you come from, whatever you do in your life, you're always busy telling a story. . . . It's a continual thing, this telling stories" (Profeit-LeBlanc, 2002, p. 47). Given this continual storytelling, all research is a form of storytelling, even quantitative research (Clough, 2002). However, portraiture is particularly interested in storied descriptions of experience and phenomena, and it uses a literary, story-telling form to share knowledge. These choices carry assumptions about valuable knowledge and enable particular kinds of knowledge to emerge.

Identifying storytelling as a location for theorizing draws from feminist and narrative research approaches. These approaches assert the value of lived experience as a place from which to generate knowledge (Fawcett, 2000; Gaard, 2009; Gough, 1999; Hallen, 2000; Harvester & Blenkinsop, 2010; Richardson, 2000). That is, these approaches suggest that the stories people tell to make sense of their lives are the foundation for theorizing at larger socio-cultural levels. Storytelling as a way of theorizing enables people to produce knowledge that is clearly tied to context, community, and that is co-constructed. Richardson (2000) describes how feminists in particular have moved away from understanding theory as architecture, and toward the idea of theory as story. She writes,

> In the 1970's, feminist researchers introduced and acted upon a different metaphor: "Theory is story." Not only is the personal the political, the personal is the grounding for theory. With the new metaphor for their work, many feminists altered their research and writing practices; women talking about their experience, narrativizing their lives, telling individual and collective stories became understood as women theorizing their lives. (p. 927)

Conceiving of storytelling as a kind of theorizing ties back to the point that Lawrence-Lightfoot and Hoffmann Davis (1997) make (explained above) about how portraiture finds the general in the particular. For example, in the excerpt from Ray Barnhardt's portrait above, there is ample opportunity to theorize about positionality based on the stories that I shared of my own feelings of insecurity and Ray's politics in a First Alaskan educational context.

In addition to portraiture valuing storytelling as a location for knowledge creation, portraiture's use of story enables a kind of knowledge to emerge that is contextual, co-created, complex, and aesthetic. Researchers that use narrative inquiry readily acknowledge the temporal, place-based, and social nature of stories (Clandinin, Pushor & Orr, 2007). Similarly, portraiture recognizes that knowledge cannot be separated from the people, places, and times with which it was created, and thus often does not hide, remove, or generalize the particularities of people and places (Cheney, 1989). In these ways, portraits push back against the tendency to compartmentalize knowledge in their often-blurred lines between the personal and professional, or the public and private. Thus, portraiture's use of story as a way to share knowledge is less reductionistic than some other methodological approaches. Following the thread of a story, researchers weave their way through categories that might otherwise be used to cut up themes into seemingly discrete pieces.

Additionally, because portraiture listens for story, researchers are given the opportunity to consider multiple storytellers. Lawrence-Lightfoot and Hoffmann Davis (1997) explain a distinction between listening *to* a story, a more passive approach where the researcher "waits to absorb the information and does little to give it shape and form" (p. 12), and listening *for* a story, a more active approach where the researcher is "more active, engaged" (p. 12) during interviews, asking questions that share "the language of the narrative that *we* are co-constructing" (p. 121). When researchers listen *for* stories, they do not attempt to blend in to the background, they acknowledge their role in shaping the discussion, identifying the story as co-created.

While Lawrence-Lightfoot and Hoffmann Davis (1997) discuss the co-creation of stories as a process by which the portraitist's and participant's voices work together to co-construct a story, my experience is that the variety of voices involved in the co-creation is much larger. The stories we tell have their origins in a complex milieu of voices coming together at a given point in time. For example, the short story from Ray's portrait above includes the voices of his mentors, past students, a particularly effective virus, myself as a researcher, and a current political narrative of our time. The story is co-created in this complex time, place, and mix of voices. Our stories are continually informed by these types of varied voices (and more!); portraiture's blend of ethnographic and narrative approaches to research allow researchers to seek out the voices of more than just the participants themselves. This can mean interviewing participants' peers, attending to one's own voice as a researcher, and it could mean recognizing voice in the more-than-human as well (Timmerman & Piersol, 2013).

Multifaceted conceptions of voice have often been used in research to narrow in on meaning, triangulating between different voices to find the one kernel of truth. However, in portraiture, multifaceted conceptions of voice also increase the ambiguity of experience and expertise. In a context where it is easy to oversimplify, portraiture's inclusion of multifaceted voices helps preserve the complexity, co-creative, and relational nature of our experiences and ideas.

Finally, portraiture's use of story and literary devices also produces an aesthetic form of knowledge. Lawrence-Lightfoot and Hoffmann Davis (1997) explain that portraits should be literary, artistic, and accessible, using narrative to express "attitudes, feelings, colors, pace, and ambiance"

(p. 28). These aesthetic expressions "do more than refer... to the object of representation, their own properties have significance in themselves" (ibid). Employing metaphor, variations in tone, and intentional use of flow in the portrait gives researchers the ability to express a wider array of information. For example, concepts or experiences that might be otherwise difficult to quantify or evaluate, or are simply ineffable, can be brought to life through story. Their weight, intention, and meaning can be conveyed in the simple description of the brush of a hand, a pause, or a tear rather than attempting to find narrow words for the actual phenomena itself. That is, the aesthetic aspects of portraiture are not merely window-dressing, but are meaningful in and of themselves.

Further, portraiture's use of literary storytelling asserts the value of aesthetic and subjective knowledge. Unashamedly including, and even foregrounding, the otherwise commonly marginalized "evocative, emotional, nonrational, subjective, metaphoric" (Richardson, 1997, p. 39), and sensory qualities of experience, makes an ethical statement about the value of these experiences. "In performing our subjectivities, we assert the relevance, the legitimacy, indeed the necessity of including the full range of our humanness in our work of re/membering ourselves in/to the world, embracing the world, with all our relations" (Hasebe-Ludt, Chambers, Oberg, & Leggo, 2008, p. 68).

Visions of Interventions

We are walking again. This time, we're walking up. David's hand reaches out periodically to brush by or lightly grasp the oceanspray, the ponderosa pine, snowberry, and others. Eventually emerging from the trees onto the exposed rock at the top of Kamiak Butte, we are greeted by big sky and big land. Several hawks circle above and the fields are gold with fall crops. Periodically standing, resting, listening, watching, and talking, this is where we have our last interview. The audiotape records almost as many silences as it does conversation, and is dotted with hawk cries and gusty winds throughout.

Figure 2.5. Golden wheat and ponderosa. Original drawing by Nora Timmerman

Sketches of items I gathered on my visit to the Palouse (wheat and ponderosa pine needles) reflect the dominant view from the top of Kamiak Butte where we recorded our second interview.

We talk for an hour or two, I am not particularly conscious of the time. Toward the end of our conversation, and after a lengthy pause, I recall how David's desire for social change is connected to his reading of Thoreau, and I ask him, "What does the civil disobedience of an academic look like? . . . Not just an academic, but a place-based academic?" Another long pause follows as he mulls the question over. Ideas and answers are tossed about over the next few minutes, and David's thoughts seem to trace two paths. The first has to do with "taking some form of direct action, . . . not buying in to the productivity discourse," saying " 'No, I'm not going to do that,' and being willing to face consequences." However, he explains, "There's civil disobedience, and then there's being an agent of change –and sometimes they overlap, and sometimes they're separate projects."

Returning to Thoreau, David maps out the ways in which writing—a common "outcome" of the academic push for productivity, and the second path he discusses—can actually be a powerful source of social change. "I think that's one way, is in writing, which is of course, what Thoreau's essay was. His act meant nothing, it was the writing of it that became a movement, and one that's been taken up by lots of groups on the ground in lots of different places, so there's definitely value in that." And, ironically, "The good thing about academe is that if you couch your discourse in academic language, you can be pretty radical. . . . I've been very critical of the very institution that pays my bills . . . in an effort to help create an environment that feels more sustainable, more in tune with people, place, and land."

While David has not written a book that has received attention in the way Thoreau's Civil Disobedience has, the 2003 "Critical Pedagogy of Place" article he wrote for Educational Researcher continues to be a reference point for many scholars—critics and fans alike. Indeed, David's publishing record is substantial, not only in terms of its volume, but in terms of its scope and effects. Many eco/environmental educators know his name and are affected by his poignant writing. Having worked for years in public schools as an English teacher, his talent for crafting gripping essays and poetry is exceptional. Publishing ecological education pieces in mainstream education journals is thus another potentially powerful and under-utilized avenue for social change that David suggests. As of my visit with him, he said that there had not been another environmental or ecological education piece published in Educational Researcher since his in 2003, and David admits, "I don't understand why other people don't do it—I just don't get it. Or, maybe they are and these articles are just being weeded out. . . . I really think that people in environmental ed. that have this broad analysis of culture and environment really ought to reach a wider audience, be more strategic about how to make an impact on the field."

Another silence follows these thoughts. Squinting at the brightness of the sky, David points North, "More redtails in that little slice of sky." We hear their cries as they prey upon the abundant rodents in the fields below. Our final interview is circling to a close. David repeats my last question, ". . . it's what a lot of us at these EE conferences have been talking about: what should we do as a group, how do we make a difference, how do we interrupt the status quo, how do we make an intervention? . . . What should be our act of civil disobedience or our intervention at this point in time? I think that's a really good question..." And this time, the question is left as a point of reflection for the future.

Toward Ecoportraiture

For all of my discussion of portraiture, this is a book about *eco*-portraiture. Ecoportraiture is a new idea, one that our various chapters are attempting to elucidate and imagine, and one that has the potential to be a piece of what David and I discussed when we wondered how to "make a difference, interrupt the status quo, and make an intervention." Our editors write that ecoportraiture is not only a methodology for "fostering human relationships and providing insight into their embeddedness in community and place, but also for ***including the natural world as participant, actor, teacher, and collaborator***" (emphasis in original, Book Proposal). I did not use the term ecoportraiture for my work, but I was interested in "including the natural world as participant, actor, teacher, and collaborator" from the start. Despite this interest, this kind of inclusion was not a key purpose of my work, and thus my efforts toward it were not extensive or centered. In this section, I share reflections on what I did, why, and what I might do differently if I were to set out with an intention of doing ecoportraiture.

What I Did: Eco-ish Portraiture

For me, the very notion of ecoportraiture affirms moral concern for the natural world (Moore, 2005). Ecoportraiture for the more-than-human is (in part) portraiture done with the intent that knowledge created and shared through the research contributes to the thriving of all life now and in the future. In particular, portraiture offers the opportunity to do research for something by telling a new story about that thing. In other words, because portraiture works through and with stories, it gives researchers an opportunity to listen for and tell stories that are marginalized or silenced. While other research methodologies (critical oral history, feminist narrative inquiry, etc.) do similar things, the form of portraiture is unique among them as a methodology that results in cohesive stories (Barone, 2001; Clough, 2002). Lawrence-Lightfoot (2005) thus suggests that portraits themselves are acts of "intervention." She explains,

> In the process of creating portraits, . . . we engage in acts (implicit and explicit) of social transformation, we create opportunities for dialogue, we pursue the silences, and in the process, we face ethical dilemmas and a great moral responsibility. This is provocative work that can disturb the natural rhythms of social reality and encounter; this is exciting work that can instigate positive and productive change. (p. 12)

Portraiture acts as a transformative intervention for participants, researchers, and readers, offering each new stories by which to make sense of and come to know our lives.

In my research, I attempted to affirm my concern and care for the natural world by inciting and inspiring readers, politically and morally. A central purpose of sharing portraits is to offer opportunities for (critical) reflection among readers. "The critical educational storyteller is out to prick the consciences of readers by inviting a reexamination of the values and interests undergirding certain discourses, practices, and institutional arrangements" (Barone, 2001, p. 193). For my specific study, the common story about integrity and contradiction is that environmental educators must seek purity and perfection in their lives to match their ethical stance. For example, many environmentalists are told (and tell themselves) that they should not own a car, have children, turn on a light, or buy clothes if they *really* cared about the environment. In response to this story, I wanted to tell new stories that show how contradiction is not avoidable and so I looked for stories that shared the dialectical nature of integrity and contradiction. I argue that these more complex and nuanced stories affirm moral concern for the more-than-human world because they orient educators away from neoliberal, individualistic action that inadvertently perpetuates some of the underlying roots of ecological destruction. Telling a story about integrity as a dialectic versus integrity as a kind of moral perfection or alignment encourages people to work together, to not be paralyzed by their complicity, and instead to be motivated and drawn on by it.

In addition to affirming the political and moral intent of my research, I also made a few small attempts to, as our editors say, "*include[e] the natural world as participant, actor, teacher, and collaborator.*" Primarily, these attempts took shape in the ways I listened for stories, analyzed my data, and shaped the portraits.

While my research portraits are ultimately stories about individual educators, I sought to understand the primacy of participants' relationships with the more than human in the portraits, asking how those relationships actively shaped participants' sense of self and conceptions/experiences of integrity. During my site visits, for example, I paid particular attention to participants' interactions with nature generally, asking myself how I see them demonstrating their spoken connection with the natural world through day-to-day actions (acknowledging, however, that these demonstrations are not always visible, especially to an outsider, such as myself).

My data collection included gathering natural and found objects (subjects?), drawing, and taking pictures. My hope for these practices was to engage with participants and participants' places at a more sensory level. I wanted to expand my ability to "know" them through more-than-verbal means. To do what ecoportraiture asks, to include the natural world as participant, actor, teacher, and collaborator, we need to find other ways of communicating beyond the English language. My use of pictures, drawing, and gathering objects from my participants' home region was a small exercise of these other, more sensory ways of communicating.

I kept any gathered items with me, on my desk, as I transcribed, analyzed and wrote up the portraits—for example, the wheat and ponderosa pine pictured in the excerpt above. I also made a practice of listening to my recorded interviews while on walks in the forest. I felt it was important to be outside while I was trying to make sense of what I learned, hoping that being in "natural spaces" would shape my interpretations through the reference points and sensory input of the more-than-human world with which I was interacting. Finally, as demonstrated in the portrait excerpts above, I included images throughout all of my portraits, and virtually all of the images in the portraits are of the natural world. These images are intended to communicate the significance of these places in participants' stories—how the place itself is a teacher and collaborator with participants.

These examples of what I did toward ecoportraiture are not particularly unusual and certainly not revolutionary. I am not convinced that they gave me "access" to something like a more-than-human voice or co-constructed knowledge beyond our standard ways of knowing (which are always, already co-constructed with the natural world, though this is generally denied). They pointed in a direction, exercised certain skills, and turned toward an acknowledgment of the subjectivity and moral affirmation of all life. Should I attempt the same research project with an overt effort to engage in ecoportraiture, I would have done things differently.

What I Might Do: Possibilities for Ecoportraiture

If I were re-do a portrait or another aspect of my research to more fully embrace the call of ecoportraiture to "include[e] the natural world as participant, actor, teacher, and collaborator," I would need to re-frame my research from the start. As a faculty member who frequently advises graduate students as to the importance of the connection between their chosen research methodology and their research question, I cannot think about what it would look like to have used ecoportraiture for my dissertation without shifting the question and purpose of the work. In my dissertation, my research question was: "How do respected educators who recognize and critique the ways in which dominant, modern, institutionalized education contributes to ecological concerns, work toward and conceptualize ecological integrity within their personal and professional lives?" And, I described the purpose as follows: "The purpose of this dissertation is to inquire into, critically explore, and share thoughtful possibilities for teaching and living with ecological integrity." To have ecoportraiture be the right methodological approach, I might have asked instead: "How has the natural world taught educators about what it means to live and work with integrity, and what did they learn?" The purpose would shift as well to something like, "the purpose of this research is to explore and share both how the natural world teaches educators about integrity and what it teaches educators about integrity."

With this new question and purpose, so many aspects of the data collection would also be different: interview questions, the locations of our interviews, and my field notes. The focus of the research would shift from the individual educators to the relationship between those educators and the more-than-human entities from whom they learn. I would ask my participants to take me to places where they have learned about integrity, and I would somehow try and engage with those entities myself, or shift my focus to searching for the relationship between the educator and the natural world... draw those or photograph them, trying to see them from my own perspective.

My analysis and representation would change with this shift in research question and purpose. The analysis would seek to tell a story about a relationship instead of a story about an educator. This would require that I analyze the different components of that relationship and how they interact. The portrait would also be less focused on a person and more focused on a process and lessons that emerge in relationship with the natural world.

Of course, I have no idea how this would work or if it would work; it is all guesswork at this point. The goal of this brief guessing here is to question my way toward ecoportraiture.

Conclusion

Portraiture has been an incredibly rewarding research methodology, and I am only at the beginning of exploring its possibilities. Through my dissertation, portraiture encouraged and led me to befriending three amazing educators. Writing portraits saw me pulling out my hair at times, like many grad students and writers, but it was absolutely a joy. Playing with adjectives, tone, structure, themes, and syntax simultaneously expressed what I had learned as well as the feeling I behind my experience.

Ecoportraiture remains an unknown possibility and a project of collective imagination for me. There are amazing examples in this book of scholars who have spent more time imagining and practicing ecoportraiture. As I continue to imagine, question, and look toward the possibilities for this methodology, I will want to do so in their company. Together, through the characteristics of relationship, complexity, co-creation, and artistry that portraiture already exemplifies, we may find something worthwhile.

References

Barone, T. (2000). *Aesthetics, politics, and educational inquiry: Essays and examples.* New York, NY: Peter Lang.

Barone, T. (2001). Science, art, and the predispositions of educational researchers. *Educational Researcher, 30*(7), 24–28.

Berry, W. (2012, April). It all turns on affection. *Jefferson Lecture in the Humanities.* Lecture conducted from The National Endowment for the Humanities, Washington, D.C. Retrieved July 27, 2013, from http://www.neh.gov/about/awards/jefferson-lecture/wendell-e-berry-lecture

Cheney, J. (1989). Postmodern environmental ethics: Ethics as bioregional narrative. *Environmental Ethics, 11*, 117–134.

Clandinin, D. J., Pushor, D., & Orr, A. M. (2007). Navigating sites for narrative inquiry. *Journal of Teacher Education, 58*(21), 21–35.

Clough, P. (2002). *Narratives and fictions in educational research.* Philadelphia, PA: Open University Press.

Eisner, E. (1997). The promise and perils of alternative forms of data representation. *Educational Researcher, 26*(6), 4–9.

English, F. W. (2000). A critical appraisal of Sara Lawrence-Lightfoot's portraiture as a method of educational research. *Educational Researcher, 29*(7), 21–26.

Fawcett, L. (2000). Ethical imagining: Ecofeminist possibilities and environmental learning. *Canadian Journal of Environmental Education, 5*, 134–149.

Gaard, G. (2009). Children's environmental literature: From ecocriticism to ecopedagogy. *Neohelicon, 36*, 321–334.

Geertz, C. (1973). *The interpretation of cultures.* New York, NY: Basic Books Inc.

Goetz, J., & LeCompte, M. (1984). *Ethnography and qualitative design in educational research.* Orlando, FL: Academic Press.

Gough, A. (1999). Recognising women in environmental education pedagogy and research. *Environmental Education Research, 5*(2), 143–161.

Greenwood, D. A. (2010b). Nature, empire, and paradox in environmental education. *Canadian Journal of Environmental Education, 15*, 9–24.

Gruenewald, D. A. (2003). The best of both worlds: A critical pedagogy of place. *Educational Researcher, 32*(4), 3–12.

Hallen, P. (2000). Ecofeminism goes bush. *Canadian Journal of Environmental Education, 5*, 150–166.

Hardt, M., & Negri, A. (2000). *Empire.* Cambridge, MA: Harvard University Press.

Hart, P. (2002). Narrative, knowing, and emerging methodologies in environmental education research: Issues of quality. *Canadian Journal of Environmental Education, 7*(2), 140–165.

Harvester, L., & Blenkinsop, S. (2010). Environmental education and ecofeminist pedagogy: Bridging the environmental and the social. *Canadian Journal of Environmental Education, 15*, 120–134.

Hasebe-Ludt, E., Chambers, C., Oberg, A., & Leggo, C. (2008). Embracing the world, eith all our relations: Métissage as an artful braiding. In S. Springgay, R. Irwin, C. Leggo, & P. Gouzouasis (Eds.), *Being with A/r/tography* (pp. 57–68). Rotterdam, The Netherlands: Sense Publishers.

Kiessling, K. (1999). Victim of corporation greed: The mysterious tombstone in Palouse. *Bunchgrass Historian, 25*(1), 22–24.

Lawrence-Lightfoot, S., & Hoffmann Davis, J. (1997). *The art and science of portraiture*. San Francisco, CA: Jossey-Bass.

McKenzie, M. (2006). Three portraits of resistance: The (un)making of Canadian students. *Canadian Journal of Education*, *29*(1), 199–222.

Moore, J. (2005). Barriers and pathways to creating sustainability education programs: Policy rhetoric and reality. *Environmental Education Research*, *11*(5), 537–555.

Profeit-LeBlanc, L. (2002). Four faces of story. *Canadian Journal of Environmental Education*, *7*(2), 47–53.

Richardson, L. (1997). *Fields of play: Constructing an academic life*. New Brunswick, NJ: Rutgers University Press.

Richardson, L. (2000). Writing: A method of inquiry. In N. K. Denzin & Y. S. Lincoln (Eds.), *Handbook of qualitative research methods* (pp. 923–948). Thousand Oaks, CA: Sage Publications.

Timmerman, N. (2013). *Coherence, consistency, contradictions: Portraits of post-secondary educators seeking integrity*. Unpublished Doctoral Dissertation, UBC.

Timmerman, N., & Piersol, L. (2013, June). *Storied sketches of people and place: Portraiture in environmental education research*. Paper presented at the Canadian Society for Studies of Education, Victoria, BC.

3. *Getting to Goodness in a Green Schoolyard: Trusting the Process of Ecoportraiture*

Kelly E. Keena

Introduction

> *The day started as serene. The Habitat is covered by six inches of fresh snow and crusted in place by the below freezing temperatures of the past few days. It was that white blanket described poetically in novels and opening scenes of movie scripts. A few animal tracks cross-hatched the snow, otherwise the view of the landscape from the warm library windows was sparkling, white, peaceful, quiet, and still.*
>
> *Outside the teacher walks directly to her place to draw. The students stream into The Habitat around her. Everything is silent. The children are reverent. Reverence is the best word that I have to describe the scene. Every child I watch moves slowly with curiosity. (Keena, Impressionistic Record)*

In my early career, I spent two years as a naturalist where my work was purely observational. I observed behaviors of birds, insects, and mammals like elk; I took copious amounts of photographs so I would remember common and scientific names and carried a journal with me everywhere to record phenology using my favorite pen, watercolors, pastels, and colored pencils. I spent one beautiful early summer morning lying on my belly next to a dear friend and colleague while we watched digger bees emerge from their straw-width tunnels on the edge of the trail in the foothills of Denver and fly back in to clear out soil. I remember the feeling of science meeting magic as I realized the circumference of the tunnel matched exactly that of the bee's body, and felt nothing but awe as the bees flew in head first, only to surface seconds later, up-side-down, the wings on their backs acting as shovels to push fallen grains of dirt out of the tunnel. I felt the same wonder and amazement when I worked with students in natural areas in my career as an environmental educator and K-8 science teacher. I felt awe listening to their dialog or seeing their faces when they first stepped into the icy cold river on a hot day.

In an elementary school's small outdoor classroom/native landscape, the scene and sense of reverence described above played out time and time again. As the science teacher in this school, this place called The Habitat was my classroom. The trail that led me to research originated in my feelings of wonder and awe regarding children's experiences in nature over the previous fifteen years. I felt compelled to understand what was happening and to explain why these encounters appeared to be so powerful.

The subject of this eco-portrait was the outdoor classroom of a public school south of Denver, Colorado, which was also certified as a Schoolyard Habitat by the National Wildlife Federation (NWF). This quarter-of-an-acre parcel connected to the school was designed, built, and used by the students following the construction of a library addition that damaged the semi-native habitat behind the school. In four years of teaching in this natural setting, I observed questioning and engagement in my students, and wanted to explore what they were experiencing. This focus on children's experience was inspired by Hart's seminal study on children's places (1979); I defined "experience" in my own study as what youth in fourth, fifth, and sixth grades (ages 11–14 years) knew, how they felt, and their actions in a natural setting. Hart's work helped me understand how to be in the space observing the students in a way similar to my work as a naturalist – to be still and discover the relationships being forged within and with this place and between beings therein. My discovery of Lawrence-Lightfoot's (1983, 2005) method of portraiture helped shape my approach to this inquiry, showing me how to be included as their teacher and co-create the study *with* them.

I entered educational research thinking I needed a problem to identify, a pathology to fix. Instead, I found a method searching for goodness that represented the richness of students' lived experiences. It was ecoportraiture that not only allowed me to follow my great curiosity with regards to how children interacted with the natural world but also allowed me to bring my multiple identities of teacher, parent, naturalist, scientist, and geographer to the work. In this chapter, I frame my search for goodness of learning in an outdoor setting by explaining why I chose ecoportraiture, the data sources that allowed the students to co-construct the project, and how they shared in the analysis process. I will end by reflecting on why ecoportraiture is a valuable research method for use in schools and outdoor learning environments.

The Search for Goodness

Lawrence-Lightfoot's concept of holistic inquiry through a lens of goodness drew my attention to her method. She initially defined portraiture while working in schools, as looking for what was working rather than what was not (Lawrence-Lightfoot, 1983). As a public educator, I too wanted to understand goodness rather than the pathologies that were the focus of most interventions and inquiries. Her approach also allowed for the presence of the researcher in the portrait, a critical piece for the teacher in the space as the researcher, as well as the co-construction of the voice and aesthetic whole by the participants in the setting.

> Goodness is a much more complicated notion that refers to what some social scientists describe as the school's ethos, not discrete additive elements. It refers to the mixture of parts that produce a whole. The whole includes people, structures, relationships, ideology, goals, intellectual substance, motivation, and will. It includes measurable indices such as attendance records, truancy rates, vandalism to property, percentages going on to college. But it also encompasses less tangible, more elusive qualities that can only be discerned through close, vivid description, through subtle nuances, through detailed narratives that reveal the sustaining values of an institution. (Lawrence-Lightfoot, 1983, p. 23)

Lawrence-Lightfoot was inspired by the process of having her portrait painted, thus the name of the method. She described the portraiture process as revealing the essence of the portrait's subjects in a way similar to her experience as the subject of a painted portrait in the traditional sense: "I wanted the subjects to feel seen like I had felt seen – fully attended to, recognized, appreciated, respected, and scrutinized" (Lawrence-Lightfoot, 2005, p. 6). This work of art depicting a person became a work of science to create depiction of a place comprised of many people. This method allowed the dynamic nature of schools to emerge, with all their complexities and nuances, rather than focus singularly, or myopically, on an inquiry into schooling in a narrower sense.

Ecoportraiture builds on Lawrence-Lightfoot's method, an evolution allowing for the complexities of human systems in natural or green spaces. This inquiry approach is actually rooted in the complexities of dynamic natural systems to illuminate the other-than-human and human intersections of the landscape, biotic and abiotic. In the context of ecology, ecoportraiture explores the goodness of these interactions and creates a portrait of the whole, not the parts.

This conception of ecoportraiture emerged in the context of my observations of students' interactions in and with The Habitat, an outdoor natural space connected to the school and rehabilitated by the students and teachers based on the NWF Schoolyard Habitat program's guide. The Habitat included some cultivated sections (a butterfly garden) and some spaces left feral. The quarter-acre space was naturally surrounded by Gambel Oak, a predominant native shrub that created a wall of woods with several planted Ponderosa Pine, also native to the region. A break in the scrub oak led down a hill to a drainage that was technically not part of The Habitat, but became an extension of the place and was used for classes. At the bottom of the hill was a seasonal stream that connected to the town's main water way. In both the drainage and The Habitat, it was common to find evidence of black bears, coyotes, mule deer, foxes, raccoons, and a cacophony of native and migratory birds. The Habitat was designed for the school as well as the community, and so there were purposefully no fences or walls keeping children out or in. Many students walked through the drainage to get to their homes and many children walked to The Habitat as a community space.

One quality of children's relationships to nature that ecoportraiture serves well to illuminate is the social dimensions of all natural places. Ecoportraiture includes the complex relationships within a place, between and among *all* the biotic/abiotic players in an ecosystem *including* people. For example, the development of The Habitat included the natural filling in of a water drainage, which eventually became a thriving aquatic system, or what the kids simply called "the pond". This became a hub of activity and investigation, but also drew in the wildlife. Through the multiple data-gathering methods employed, the students' varied experiences of the pond and its teeming natural processes appeared in the final eco-portrait.

This method of research also allowed me to sit quietly and observe children as they moved through their schoolyard, to ask them what they were thinking about, and to explore what motivated their movements and inspired their choices. Ecoportraiture was the perfect method to understand my students' experiences. Understanding what children knew, how they acted, what they did, where all this took place and with whom, was gathered through a catalog of individual data sources that, when viewed together, created the whole eco-portrait of this place.

How is Ecoportraiture Different?

Ecoportraiture's roots in phenomenology and the tradition of hermeneutics created a method well-suited for studies of schools and outdoor learning environments. Cresswell (2007) describes phenomenological study as a method that "describes the meaning of several individuals of their lived experiences of a concept or a phenomenon" (p. 58). Lawrence-Lightfoot and Davis (1997) describe portraiture as a "method of qualitative research that blurs the boundaries of aesthetics and empiricism in an effort to capture the complexity, dynamics, and subtlety of human experience and organizational life" (p. xv). The eco-portrait of The Habitat was a portrait of a place experienced by a variety of actors, rather than a portrait of people in a setting. The eco-portrait of The Habitat was an artistic account *of* a particular setting comprised by a group or culture, guided by questions not about who the children were, but who they were in the context of a very specific, geographically, and temporally bounded setting. It was the stories of many children (n = 106) that provided the rich texture and detail illuminating the place itself. To create the final portrait of this place, multiple forms of data were incorporated into the research design. Each source informed an iterative and generative process of understanding, allowing the flexible emergence of the resulting eco-portrait depending on what the students reported or revealed.

Where Gruenewald (2003) asks us to learn from places by asking what happened here, Lawrence-Lightfoot (1997) encourages empirical portraits to discover "What is happening here, what is working, and why?" (p. 142). Both ask us to probe deeply into places and the human relationships with and in them, because inhabitants of places are inextricably linked to those places. Few places exemplify that more than a school, where children and their teachers spend around 175 days a year for more than seven hours a day. A study of a school must include the human systems and structures that make up the daily functions of the school. Ecoportraiture afforded a method of inquiry that honored the schooling system but explored what happens when students and their teachers step outside into a natural setting during the school day. I came to call this understanding, this connection, "how we dwell"—how we as humans inhabit our places and nestle into the human systems therein, but also how we live with and amongst the natural aspects of our daily settings.

Other dimensions of ecoportraiture lending value to the study of learning and outdoor environments include the allowance for the researcher's voice, co-construction of data with participants, and rich descriptions. The above definition of our dwellings, or contexts, contains a temporal element, historical and present, and includes the portraitist's relationship to the place and her personal story there. This was critically important for me, as a teacher who noticed a social/cultural/learning phenomenon in a natural/ecologically rich classroom and wanted a way to explore, define, and shine a light on what it was. With more efforts to welcome in the teacher voice, potentially we could see more teachers leading research in their schools, places where *they* dwell.

The inclusion of my voice was also important because I developed and implemented many of the rules and rituals in The Habitat. Those that ended up being important in the study were "Two minutes of silence" and "Wonder Days". Two minutes of silence was a ritual I created to help the students separate the time in The Habitat from recess. When the classes came into The Habitat, they were asked to sit on the steps next to the pond for two minutes, silently, before getting instructions for the day or before setting out to investigate. Wonder Days were a ritual I created in response to a lot of structured lessons in The Habitat, as teachers in the school learned to balance student- and teacher-directed time in an outdoor space. Wonder Days were an agreement I had with the students: they would stay in the boundaries of The Habitat and I would not direct them on where to look, what to look for, or how to look for it. It was an unstructured time for students to explore during science class. Both of these activities allowed the students to settle into place and begin to attend, experience, even dwell.

As in phenomenological inquiry, ecoportraiture seeks to find the essence of the experience; ecoportraiture, also allows the actors in the story to become co-creators of the narrative. The co-construction of the portrait through rich data sources was based on building relationships with the participants, known as the actors within the setting. The relationships were forged through empathetic regard between teacher and student as advised by Lawrence-Lightfoot and Davis (1997) and Freeman and Mathison (2009), who also discuss the concept of goodness: "a belief in the intelligence, worth, and capability of each student is a must when working with children" (p. 84). In addition to the allowance for goodness, empathetic relationships created trust and insight. Davis explains, "empathy is seen as the channel of emotional resonance, the vehicle for gaining a deep understanding" (Lawrence-Lightfoot & Davis, 1997, p. 147).

The search for goodness fosters empathetic regard in the portraitist for her actors. This removes the risk for the student, as the storyteller, of being judged, blamed, and/or punished and allows the researcher to sense the experience from the actor's view. The data sources, explained in detail later in the chapter, were intentionally selected to provide insight from the students' perspective beyond question-and-answer type methods. At the time of the study, children's experiences were understood in this way through drawings and maps (Freeman & Mathison, 2009; Sobel, 1990), photographs (Freeman & Mathison, 2009; Rudkin & Davis, 2007), video (Carpiano, 2009; Kusenbach, 2003), interviews individually and in groups, and through conversations that accompanied drawings, maps and photographs (Christensen & James, 2000; Freeman & Mathison, 2009; Westcott & Littleton, 2005).

Now that I have presented my rationale for choosing ecoportraiture, the next section will lay out the overall study at a macroscopic level to provide context for the data sources and why they worked so well in this setting. I will also present the analysis of the data as developed through an "iterative and generative process" (Lawrence-Lightfoot & Davis, 1997, p. 185), which continues to be the most challenging and by far the most rewarding part of this approach.

The key question guiding The Habitat's eco-portrait was how students experienced their time in the setting during the school day. The following section defines my theoretical framework in ecological psychology, followed by the data sources and participant selection for the study.

Ecological Psychology, Affordances, and a Lens for Data Collection

Theoretically, ecological psychology and perceptual learning (Gibson, 1969) emphasize that each experience in an environment is unique. This is because in a setting with limitless information, like The Habitat, what information each actor selects and attends to is also unique (Chawla, 2007). Based on this groundwork, the guiding question of the portrait was not about one specific aspect of the experiences students were having; rather, the question was about the themes that emerged from the qualities of those experiences. This was told through their narratives, drawings, photos, videos, and interviews. Emergence allowed me to find a deeper understanding of how children saw themselves as part of something bigger than themselves, and when, or if, they isolated themselves apart from their natural surroundings. The eco-portrait became about the place specifically as seen through the varied experiences of the school's intermediate students. As Lawrence-Lightfoot suggests, "a whole... encompasses less tangible, more elusive qualities that can only be discerned through close, vivid description, through subtle nuances, through detailed narratives that reveal the sustaining values of an institution" (1983, p. 23).

Selection of information is an open-ended perceptual process. J. J. Gibson (1966) found that perception is an active, ongoing, full-sensory process. Perceptual learning is the intrinsic attention to changes in the environment and the ability to learn through our experiences and encounters in our environment (Heft, 2001). The ability for an individual to move through a space and interact with it through all sensory channels gives humans infinite ways to understand that place (Heft, 2001), based on who we are as individuals including our own constellation of backgrounds and experiences from before we arrive in the place. The selection of information invokes the core concept of 'affordances' (J.J. Gibson, 1979), the properties of an environment that are noticed and used by an individual. Affordances are the selected variables in a place that are used to suit the needs or curiosities of that individual. Affordances are a core tenant of understanding the role of nature in childhood (Chawla, 2007; Kirkby, 1989; Kytta, 2002; Malone & Tranter, 2003; Stanley, 2010).

This theoretical grounding suggests that encounters and experiences in the immediate environment, involving unlimited opportunities for selection and meaning, are necessary for human development. Such use of ecological information "unites an organism with its environment" (Reed, 1996, p. 52). This then forms the basis for a definition of the concept "sense of place," including the inseparable blend of affect and intellect developed in the course of children's direct contact with nature through exploration, discovery, and imagination (Kellert, 2002). This could be seen during our science classes: students entered The Habitat by passing from a human/teacher-constructed four-walled world to a classroom with no ceiling, one engorged with sounds and smells and textures as unique as the constellations that made up each child. Although this space was also planned and constructed by people, it was filled in over time by the biota of the surrounding area and was thus unwalled and unbound compared to the way an indoor classroom is. Using these theoretical lenses of ecopsychology and affordance, I moved towards a study that incorporated multiple data sources including participant observation, student maps and drawings, student interviews, go-along interviews, and my Impressionistic Record.

Data Collection

Data were gathered during normal class time in The Habitat or in private settings. In class time, the entire class did drawings as an activity. Drawings from participants in the study were included (i.e., drawings of participant students) and non-participating students' work was kept separate. Children who consented to participate were those observed during Wonder Days in The Habitat, although all children in that class were participating in the activity. These were days without specific formal learning objectives or lesson plans, where the students went outside with me as their science teacher and freely explored The Habitat. Interviews were scheduled and done privately. Participants were selected from an open option to participate for all fourth, fifth, and sixth grade students in the school. Each grade had three classes. Importantly, this age group has been noted as being the most formative in developing a connection to nature, as well as the least investigated (Sobel, 1990).

An eco-portrait of place is to be likened to a collage, with each child's portrait, comprised of stories, feelings, thoughts, and actions, being a part of the larger whole. When viewed together through emergent themes and stories that share resonance with one another, the individual portraits form an artistic expression of a whole: a place. Given the exploratory nature of my questions, I sought a method of inquiry that would allow for observation, conversation, and visual media – all of which allowed the students to be participants in both the data and the collection of the data. This participation in the research allowed the students to be co-constructors of the research process. I also required a method that did not erase my role as a teacher nor my relationship with the students out of the data. The most unique and powerful tool in ecoportraiture is the Impressionistic Record, the rigorous reflective process undertaken by the researcher that provides a personal, first-person running dialog throughout the research process. This was, by far, the most time intensive and most rewarding aspect of the ecoportraiture study; it provided great insight into and understanding of the themes as they began to emerge in the analysis of a massive amount of data.

The participants in the study were organized into three groups: 1) the sample (n = 106), 2) the sub-sample (n = 56 of the 106), and 3) Key Actors (n = 20 of the 56) (see Figure 3.1). The selection of participants was designed to narrow their numbers as their stories began to emerge in the study, in accordance with Miles and Huberman's (1994) explanation of sampling as "taking a smaller chunk of a larger universe" (p. 31). This approach to participant selection was borrowed from Hart (1979) and Childress (2000). For portraiture rooted in phenomenology, Onwuegbuzie and Leech (2005) and Cresswell (2007) recommended six to ten participants in the Key Actor group to allow for deep analysis. The Key Actors emerged from the broader sample throughout the data collection. All of the data were collected from October, 2010 to May, 2011.

Through the process of data collection, the children had a partnership role in the research as co-creators of the portrait. Alternative and creative data sources aided in eliciting responses from children and gave them a nonverbal way to communicate with the researcher (Danaher & Briod, 2005; Scourfield, Dicks, Drakeford, & Davies, 2006). In the following section, the data sources will be described with some detail, beginning with the largest data source: participant observations.

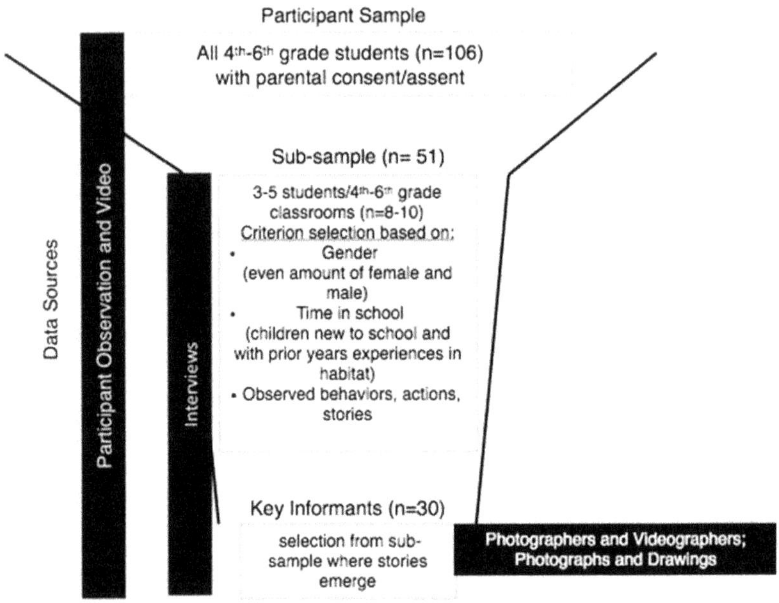

Figure 3.1. Sampling strategy and data sources

Participant Observations

Participants were observed in The Habitat during science class and during classroom time outside. Participant observations happened in two ways: 1) by accompanying the class from the classroom to The Habitat and back inside again, or 2) by sitting in one place in The Habitat (on the hill or on the rock steps) and observing multiple classes enter, use, and leave The Habitat. The second approach was useful on busy days where multiple classes were using the space. Participant observations were also spontaneous; they were conducted during class time with the students' classroom teachers. Having a rapport with the fourth, fifth, and sixth grade teachers afforded quick emails or text messages by cell phone when a class was going into The Habitat.

Participant observations evolved throughout the course of the study. Initially, a list of participants in the greater sample was referenced during class time observations to ensure the notes being recorded were from students who gave consent to participate in the study. As familiarity with participants in the sample increased, the sub-sample was selected and the use of video was incorporated. Beginning in January, the movements and behaviors of each child in the sub-sample were recorded using a base map of the habitat during observations by the researcher. This data provided a cross-reference for what children reported having visited and done during interviews. Once the Key Actors were selected, the observations focused on those children's movements, actions, behaviors, and words. Children in the sub-sample, or even the larger sample, were also peripherally observed, especially in later data analysis of video. The evolution of participant observation guided the deeper explorations of the data collection as a whole, documented in the Impressionistic Record.

The role of researcher required a way to observe the interactions of the students without being distracted by classroom management situations that sometimes arose. This balance grew steadier over time through a paradox of simultaneously being there and not being there. In October and November, the students who were accustomed to me as the person of authority would approach me with questions about going inside to use the restroom and directions about the boundaries. As time went on, the students stopped asking and relied on the classroom teacher for the information. The most difficult adjustment of teacher to researcher was for teachers interacting with me during participant observations. Colleagues first, they were accustomed to our conversations during the time in The Habitat. Video was employed to aid with this challenge and to help capture more of the experiences of the students.

Drawings and Writing

Three types of drawings and writing were collected as data: 1) independent observations in science notebooks, 2) memory maps, and 3) event maps. These three were chosen because they offered a diversity of data as well as aligning with the curricular aims of the teachers. Every student at Carson Elementary kept a science notebook as a place for recording science investigations, research, and studies in the classroom and The Habitat. During Wonder Days, all students had the choice of recording observations by writing and drawing in their science notebook. The drawings were analyzed for patterns and used in some interviews in a similar way as photos. Drawings were scanned electronically from participants after interviews when a child referred to his or her science notebook. The second form of drawing data was a memory map of The Habitat. In the science classroom, every student was asked to draw The Habitat from memory, including all of the features they could recall. Students were asked to place a star on the map indicating where something memorable happened, and a smiley face on their favorite place in The Habitat. On the back of the map, the students were asked to describe their memorable story, their favorite place, and three adjectives to describe The Habitat. Maps were analyzed for patterns and the stories, favorite places, and adjectives were recorded and placed into a spreadsheet for coding analysis. The third form of drawing data was an event map. Based on nature writer Hannah Hinchman's (1997) work, an event map is a series of sketches in a place of the events that occur over a set period of time (e.g. one drawing for every ten minutes). Students were given a base map of The Habitat that included major features as point-of-reference. Then, the students recorded drawings and words of what "events" took place as they walked around or sat in one place. Memories of events were also included. All students in each fourth, fifth, and sixth grade classroom were asked to do this activity during science specials class, except in the cases of severe weather. In those three classes the teacher agreed to do the event maps during class time. In all cases, the science teacher and the classroom teachers followed the same protocol. The drawings in students' science notebooks and the event maps were used for discussions in interviews and in some cases they were analyzed for emergent patterns in the data.

Interviews

Two types of interviews occurred: semi-structured interviews with participants in the sub-sample group and go-along interviews with Key Actors. All interviews were conducted in The Habitat, except in several instances where the weather was too severe. In the event of inclement weather, either the library that overlooked The Habitat through large windows or the science classroom was used, depending on the librarian's class schedule. Interviews took place during the students' lunch period; all interviews were recorded and transcribed. Interviews were scheduled with students as soon as the participants were selected for the sub-sample, with the goal of one interview per participant. A third-grade child in a class of mine once told me that writing about his experience was like a speed bump to his thoughts. I found that the interviews provided an opening up beyond the writing I had observed in student notebooks. In many cases, students would bring their notebooks from science class and share their drawings or notations with me.

The second interview type was a "go-along" interview based on the method defined by Carpiano (2009) and justified for this purpose by Cele (2006), Kusenbach (2003), Trell and vanHoven (2010), who all found that a walk with children in a familiar place allowed them to elicit nuances of the child's experiences and for the child to have a more active role in the research. Go-along interviews, or walking narratives, were done only with the Key Actors in the study and were held with one student at a time. Participants were given a Flip video camera and asked to hold it as they walked through

The Habitat, giving the researcher a tour of their usual activity in the space. The Flip video camera had a viewing window that framed what was being filmed without having to look through a lens; it was an appropriate tool for this type of data because the child could move through the space, viewing it normally, and still hold the camera. The participants were asked to hold the camera in the direction they were facing and up away from the ground; I followed the participant holding the camera without providing direction on where to go, what to look at, or where to stop. In many instances, I noted the child forgetting about the camera, and the narrative in the video was naturalistic and very valuable, rather than resembling a performance or being affected by nerves. These interviews were fifteen to forty-five minutes in length depending on the participant. The go-along interviews typically started with the question, "Where should we go first?" This freedom allowed for exploration of The Habitat. These walking narratives were employed with the smallest group of participants in the study and were the last phase of the data collection.

Photographs

During go-along interviews and some group interviews in The Habitat, participants were asked to photograph three scenes; the directions were, firstly, to photograph the participant's favorite place, followed by two photographs of whatever caught the participant's attention. In some cases, the photographs were used in the interviews to elicit dialog from the participants about The Habitat and to give the students a stronger voice in the research, as in cases where the subject interprets the photo for the portraitist (Harper, 1994). Rudkin and Davis (2007) used photographs in a similar way with youth in an urban area based on Wang and Burris's (1994) concept of "photovoice," when the camera gave participants an opportunity to frame their unique perspectives in the place (as cited in Rudkin & Davis, 2007). Photographs were also taken during key moments of children's interactions in The Habitat. Sometimes, in place of video, I took these photos; this was especially useful for analysis of the go-along interviews, when the children holding the video were photographed by me to provide a wider context in The Habitat's setting.

Artifacts

The faculty and parents at Carson Elementary amassed a collection of photographs and videos relevant to The Habitat and its use over six years. These artifacts, which featured students with a parent or guardian signed media release, were reviewed to validate or contest the themes that emerged from the data and to add hues of past student experiences to the final portrait, altering names and images to assure confidentiality.

Theme Development: Trust the Process

Looking back at the data after stepping away from this study, I can say with no doubt that the most valuable tool was the Impressionistic Record. Lawrence-Lightfoot and Davis (1997) provide detailed direction for this reflective undercurrent of the study and it serves as a triangulation touchstone for each individual data source as well as a running dialog of the portraitist with herself. In my study it provided a spectrum of support, from logistics, like raw data file management and notes with memory cues for the day or event, to records of my emotions stirred up by a child's story about bullying and finding solace in The Habitat, and on to notes on what to follow up with specific actors during interviews. The Impressionistic Record was a catch-all with structure and flexibility; it was an important guide throughout the study. Looking at my Impressionistic Record years later I find that I am right back in the study. I can revisit my experiences and emotions, recognize why I made the decisions I did, and see where I paused to ruminate along the way. For example:

> *What is the nature of the experiences in The Habitat? If they are "calm, beautiful, peaceful", etc....as described by the students, what does that mean? How is it that the experiences have that feel? That tenor? HOW DO I FIND A DEEPER LEVEL OF THIS NATURE OF EXPERIENCE? This will be the question that drives my selective participants who will be the Key Informants...*

Early on, it was obvious that the accumulation of data threatened to overwhelm my organizational structures and my ability to keep up. In January, three months into the study, I wrote:

> *I am riding the crest of the torrent of data.... I am staying just far enough ahead, where I feel above the water. I am worried about how close I am to being consumed by the amount of data I am accumulating. I need to be vigilant about the organization, structure, and systems I use to stay ahead.*
>
> *PROTOCOLS:*
>
> *Transcribe one week, interview one week? (not sure if this is possible yet) List all of the data that I gathered with each reflection (see 4ZNJ for example)*
>
> *Take pictures with phone and upload to private photo database/storage. Go in and rename immediately after uploading so I don't forget what they are.*
>
> *NEED PROTOCOL for student pseudonyms. Maybe use letters and numbers until I ask them for a name. (Keena, Impressionistic Record)*

This torrent of data (Lawrence-Lightfoot, 1983; Lawrence-Lightfoot & Davis, 1997) was managed daily in a spreadsheet. This data management system came out of my need to keep track of all that was going on. Eventually, I started citing pages in the Impressionistic Record where a thought was captured that I needed to call out. This illustrates the iterative nature of ecoportraiture: there was not a data collection stage separate from an analysis stage. As I collected data, I formed more questions and found themes to challenge or accept based on the next round of data. This is the "probing, layered, interpretive" (Lawrence-Lightfoot, 2005, p. 5) process referred to when Lawrence-Lightfoot and Davis (1997, p. 185) labeled the final portrait as iterative and generative.

Data analysis and interpretation began on the first day that data was collected. The daily work involved writing in the field notebook, transcribing, reflecting in the Impressionistic Record, and preparing for the next day, always working towards a coherence. Data analysis was complex; Lawrence-Lightfoot and Davis (1997) describe construction of a portrait as being like weaving a tapestry, which happens through a tedious, systematic search for patterns. To build the overarching story, the structure and form of the conjoined narrative, and a clear and consistent voice, I drew on the five ongoing steps of data analysis according to Lawrence-Lightfoot and Davis (1997, pp. 193–214).

Steps one through three worked as an interrelated system of listening carefully to the data for 1) repetitive refrains, 2) resonant metaphors, and 3) cultural rituals using pattern coding. This listening occurred through reading, coding, re-reading, and code-checking each transcription. At the beginning of the data collection, this was a cumbersome and time-consuming process that felt more like sorting shapes and sizes; yet, as the data collection process continued and I began to feel more comfortable with the process, categories of phrases and terms began to appear as metaphors and the coding became the themes. For example, one of the repetitive behaviors and discussions that was the most difficult to pin down was related to the scrub oak surrounding The Habitat. It was related to safety and inquiry and excitement, but it eluded me until I saw a reminder of Peter Pan and The Lost Boys. Suddenly, the metaphor illuminated the sense of adventure felt by the students when they entered and explored the tangled branches that made up the oaks. This showed me that the iterative coding process worked if I was willing to leave enough space for disparate things to come together.

Using the fourth step, triangulation, I sought out confirmation and disagreement in the data sources I already had, and began to look for a sense of adventure in participant observations. The data from the visual media was incredibly useful in the process of identification of themes and metaphors, but also in attempting to triangulate the themes. Another process of triangulation I employed was to look for dissonance or the "deviant voice" (Lawrence-Lightfoot & Davis, 1997, p. 193). That provided an unsettling sense that the trail leading to a theme was not a trail at all. The discordant data challenged my assumptions related to the children's experiences and forced me to go back into the data to question those assumptions.

The final step of the analysis process was the construction of the themes. In the case of the identified theme of adventure, the iterative dives into the data also turned up a parallel theme of imagination. This understanding helped in the theme development and each of the related categories of data, or buckets of similar phrases, patterns, and metaphors, was found to have a parallel thematic partner.

In sum, the five themes of experiences that emerged from this study were: 1) critical thinking and curiosity; 2) ownership and identity; 3) peace and calm; 4) respite and respect; 5) adventure and imagination. Once these five themes were firmly established after a final review and coding process at the end of the study, relevant dimensions that supported each theme were outlined including a dimension of dissonance, or deviant voice. The relevant dimensions were filled in by stories and experiences in the data that provided evidence for the theme. Next, for each theme, a series of "If... then..." statements was created and tested by the data in the respective dimensions of the narrative (Miles & Huberman, 1994). After multiple tests of the "If... then..." statements, the qualifying stories filled in the dimensions of the themes and the themes became the coherent final narrative called the aesthetic whole. For example, a statement that was tested in this way was *if* classes spent more time in The Habitat, *then* the students' attention was directed to smaller details. Using that statement, evidence was gathered to test whether the statement was true or false. In this example, the statement was supported by the evidence and became a resonant aspect of the theme titled Critical Thinking and Curiosity. In portraiture, the data analysis process is called illumination, where many parts assemble a portrait that "resounds with authenticity" (Lawrence-Lightfoot & Davis, 1997, p. 247). This authenticity provokes a "yes, of course" response from the audience. In this case, I presented the final themes and supporting evidence to the teachers in the school to look for this response. If "yes, of course" was not forthcoming, this was a helpful signal that I should return to the data and search for other themes and/or interpretations.

The process I adore in ecoportraiture, my gateway to a love of research with children and youth, is the iterative nature of the inquiry. For me, the process of investigation, of inquiry into a place, is not a fixed line with a start and finish. Rather, it is a messy circle of questioning, observing, interviewing and asking, leading only to more questions. Each time around the cycle of inquiry, I shave off another layer, not with definitive answers but with a filling in of texture and light. This all transpired through a rigorous discipline in the research process, and through the rich narratives co-created by The Habitat's human inhabitants and their relationships with the biotic and abiotic features in it. Through trusting the process and staying ahead of the mass of data, the iterations were generative and illuminating.

Advocating for More Portraits: The Sum of Many is the Whole

A friend recently said to me, "Sometimes to find goodness, you have to look at the ordinary." As portraiture is a methodology that searches for what is right, as opposed to what is wrong, his statement resonated in the final portrait of the Habitat. What appeared to be ordinary experiences of children outside were infused with goodness, deep learning and connections, and a sense of place. Children in the school's intermediate grades found play, imagination, and nature, all ordinary things that have slipped away from childhood. The culture of The Habitat was one of kindness, respect, and often, joy. In an age where schools are bully-proofing, rewarding positive behavior through prescribed programs, and locking exterior doors out of fear based on school violence events, finding a place where kindness and acceptance was the norm was a refreshing discovery.

Furthermore, as solo experiences were nourished by social experiences, the school also benefited. The individual child was nourished by the community, and the community was nourished by the interactions between and among the children. Both could only benefit from the place symbolizing kindness and affection. In this sense, The Habitat, or natural schoolyard, advocated for children by creating a place for them to *be* children. And for me as a researcher, ecoportraiture allowed for a more authentic me, in my multiple identities, to be present. This, in turn, allowed the work to honor the complexity of teaching and the active and important role the place itself and the other-than-humans therein actually play in learning, teaching, and becoming.

My friend and colleague Emily Stanley once said that it is not the *numbers* we need to focus on, it is the *number of stories* that will help us affirm that the time children spend in nature is important. Ecoportraiture provides a method that casts a net for the richness of stories from the participants in an outdoor setting. As Lawrence-Lightfoot argued, the quantities of experiences and outputs matter *alongside* the qualities, the nuances, the richness, and the vivid descriptions. Just as individual studies can be compared through cross-case analysis, multi-voiced portraits of individual places can be compared for resonance and common themes (Chalwa, Keena, Pevec, & Stanley, 2014). The use of ecoportraiture to inquire into children's experiences of place could contribute to a common discourse from multiple fields of study including formal schooling, geography, planning and development, children's wellness, ecological psychology, and environmental education. Individual portraits of children's experiences in outdoor settings could be explored for common resonance and themes and then serve to inspire and further the global dialog about children's natural places and their experiences with/in them. This opportunity to illuminate children's knowledge of, actions in and feelings about outdoor places is especially relevant to studies of children in nature.

Ecoportraiture is a powerful method that simultaneously honors the uniqueness of each individual encounter and creates a final aesthetic whole of the place by weaving a tapestry of stories together based on common themes, repetitive refrains, and the rigorous daily practice of listening for the story. Instead of attempting to create a portrait of individual people in a setting, the study of children's experiences in the Habitat was a portrait of a place experienced by a variety of actors. Research portraits can create an artistic and empirical account of an individual person or groups of people acting *in* a particular culture or setting. Ecoportraiture, though, has implications for studies of children's experiences of place because it creates an artistic account of a particular active, agential, multi-beinged setting, thereby providing a sense not of a world of detached individual learners but of a group or even culture of learning and relating.

The questions guiding my investigation were not about who the children were, but who they were in the context of a very specific, geographically and temporally bounded setting. Their feelings, actions, and knowledge were understood through multiple co-created data sources in the context of the Habitat; collectively, they assembled an aesthetic portrait of the Habitat. In other words, the stories of many children in context and relationship became the rich texture and detail of a specific place to illuminate the place itself.

References

Carpiano, R. M. (2009). Come take a walk with me: The "Go-Along" interview as a novel method for studying the implications of place for health and well-being. *Health and Place*, *15*, 263–272.

Cele, S. (2006). *Communicating place: Methods for understanding children's experience of place*. Stockholm, SW: Almqvist & Wiksell International.

Chawla, L. (2007). Childhood experiences associated with care for the natural world: A theoretical framework for empirical results. *Children, Youth and Environments*, *17*(4), 144–170.

Chawla, L., Keena, K., Pevec, I., & Stanley, E. (2014). Green schoolyards as havens from stress and resources for resilience in childhood and adolescence. *Health & Place*, *28*, 1–13.

Childress, H. (2000). *Landscapes of betrayal, landscapes of joy*. New York, NY: State of New York Press.

Creswell, J. W. (2007). *Qualitative inquiry and research design: Choosing among five approaches*. Thousand Oaks, CA: Sage Publications, Inc.

Danaher, T., & Briod, M. (2005). Phenomenological approaches to research with children. In S. Greene & D. Hogan (Eds.), *Researching children's experience: Approaches and methods*. London, UK: Sage Publications.

Freeman, M., & Mathison, S. (2009). *Researching children's experiences*. New York, NY: The Guilford Press.

Gibson, E. J. (1969). *Principles of perceptual learning and development*. New York, NY: Appleton Century Crofts.

Gibson, E. J., & PIck, A. D. (2000). *An ecological approach to perceptual learning and development*. New York, NY: Oxford University Press.

Gibson, J. J. (1966). *The senses considered as perceptual systems*. Boston, MA: Houghton-Mifflin.

Gibson, J. J. (1979). *The ecological approach to visual perception*. Boston, MA: Houghton-Mifflin.

Gruenewald, D. A. (2003). Foundations of place: A multi-disciplinary framework for place-conscious education. *American Educational Research Journal*, *40*(3), 619–654.

Harper, D. (1994). On the authority of the image: Visual methods at a crossroads. In N. K. Denzin & Y. S. Lincoln (Eds.), *Handbook of qualitative research* (pp. 403–412). Thousand Oaks, CA: Sage Publications.

Hart, R. (1979). *Children's experience of place*. New York: Irvington Publishers, Inc.

Heft, H. (2001). *Ecological psychology in context: James Gibson, Roger Barker, and the legacy of William James's radical empiricism*. Mahwah, NJ: Lawrence Erlbaum Associates.

Hinchman, H. (1997). *A trail through leaves: The journal as a path to place*. New York, NY: W. W. Horton.

Kellert, S. R. (2002). Experiencing nature: Affective, cognitive, and evaluative development in children. In J. Peter, H. Kahn & S. R. Kellert (Eds.), *Children and nature: Psychological, sociocultural, and evolutionary investigations* (pp. 117–151). Cambridge, MA: The MIT Press.

Kirkby, M. (1989). Nature as refuge in children's environments. *Children's Environments Quarterly*, *6*(1), 7–12.

Kusenbach, M. (2003). Street phenomenology: The go-along as ethnographic research tool. *Ethnography*, *4*(3), 455–485.

Kytta, M. (2002). Affordances of children's environments in the context of cities, small towns, suburbs and rural villages in Finland and Belarus. *Journal of Environmental Psychology*, *22*, 109–123.

Lawrence-Lightfoot, S. (1983). *The good high school.* New York, NY: Basic Books.

Lawrence-Lightfoot, S. (2005). Reflections on portraiture: A dialogue between art and science. *Qualitative Inquiry*, *11*(1), 3–15.

Lawrence-Lightfoot, S., & Davis, J. H. (1997). *The art and science of portraiture.* San Francisco, CA: Jossey-Bass.

Malone, K., & Tranter, P. (2003). School grounds as sites for learning: Making the most of environmental opportunities. *Environmental Education Research*, *9*(3), 283–303.

Miles, M. B., & Huberman, A. M. (1994). *Qualitative data analysis: An expanded sourcebook* (2nd ed.). Thousand Oaks, CA: Sage Publications.

Onwuegbuzie, A. J., & Leech, N. L. (2005). The role of sampling in qualitative research. *Academic Exchange Quarterly*, *9*(3), 280–284.

Reed, E. S. (1996). *Encountering the world: Toward an ecological psychology.* New York, NY: Oxford University Press.

Scourfield, J., Dicks, B., Drakeford, M., & Davies, A. (2006). *Children, place and identity: Nation and locality in middle childhood.* Trowbridge, Wiltshire: The Cromwell Press.

Sobel, D. (1990). A place in the world: Adults' memories of childhood's special places. *Children's Environments Quarterly*, *7*(4), 5–12.

Trell, E.-M., & vanHoven, B. (2010). Making sense of place: Exploring creative and (inter)active research methods with young people. *Fennia 188*, *1*, 91–104.

4. *Developing Ecoportraiture: Lessons from the Caddisfly and other Beings*

Laura Piersol

Introduction: Why Portraiture?

In the sticky heat of August I plod along the pebble beach near my house. The large rocks there jut awkwardly into your feet and make walking clumsy so I usually cross the beach by walking on the logs along the shoreline. I like to test my balance on the variety of surfaces that they present: some of the logs, like the arbutus, are long, smooth and slippery, while others are short, with uneven bark and limbs reaching up into the sky, meaning that you have to bend your body around the branch if you want to pass by it. I was attempting such a manoeuvre one day last summer when I slid off the log toward the nearby grassy bank. In front of me where I landed I noticed a black and white rope tucked into the grass. I reached down to tighten the strap on my sandal but soon flinched back, startled to see the rope *move.* I squinted to look closely and realized it was actually a garter snake in the process of shedding its skin. I sat nearby, listening to the soft scratch as it slid its sides alongside the grass, finally peeling off the whole skin intact.

What would it mean to think of this snake as an active voice/teacher in my life? It would mean acknowledging that our reality shapeshifts according to our perception; my world is very different depending on whether I see the snake as just another creature along my walking path or as a slithering teacher. For me, this snake helps me see that molting can aid growth. When looking at my own research, I think of it as helping to remove the scales over my eyes so I might see new ways forward, by including the voice of the more-than-human world in the methodology itself. The snake also inspires thoughts and ways of being, that once lay still and lifeless, to begin to squirm and seek the warmth and light of day. I wonder whether, by learning to shed presuppositions of domination over nature and then transforming our practice as educators to listen to and include the more than human world as a co-teacher, we might crack open new possibilities in our perceptive field and in that of our students.

With this in mind, I was particularly drawn to the methodology of "portraiture," which combines the widely recognized methods of ethnography and narrative inquiry. Sara Lawrence-Lightfoot (2005) describes it as "painting with words" (p.6). Lawrence-Lightfoot and Hoffman Davis (1997) particularly emphasize the importance of context in the methodology of portraiture,

> Portraitists... rather than viewing context as a source of distortion, they see it as a resource for understanding. The narrative, then, is always embedded in a particular context, including physical settings, cultural rituals, norms and values, and historical periods. (p. 12)

Thus, this type of methodology was appealing to me as I saw it as a way to highlight the layers of particularity, relation, paradox and complexity involved in human relations with the more-than-human world that was at the heart of my research. I hoped it could allow for a deeper, more nuanced understanding and expression of this relationship.

The methodology of portraiture is also described as "acts of intervention," implicit and explicit points of social transformation where "we create opportunities for dialogue, we pursue the silence, and in the process, we face ethical dilemmas and a great moral responsibility. This is provocative work that can disturb the natural rhythms of social reality and encounter." (Lawrence-Lightfoot & Davis, 1997, p.11). Each translation is a chance for perceptual shapeshifting where every portrait has the ability to introduce a perspective that was not previously apparent. Lawrence-Lightfoot (2005) describes the process of having her portrait taken early in her life:

> [I]t seemed to capture my "essence"; qualities of character and history, some of which I was unaware, some of which I resisted mightily, some of which felt deeply familiar. But the translation of the image was anything but literal. It was probing, layered and interpretive. In addition to portraying my image, the piece expressed the perspective of the artist and was shaped by the evolving relationship between the artist and me. I was never treated or seen as object but always as a person of strength and vulnerability, beauty and imperfection, mystery and openness. The artist needed to be vigilant in capturing the image but always watchful of my feelings, perspective and experience. (p.5)

"Probing, layered and interpretive... evolving," these are some of the principles she carries into her educational research methodology, aptly called "portraiture." Lawrence-Lightfoot and Davis (1997), in describing this methodology, refer to the fact that historically many artists and scientists were aware of "their inability to capture and present the total reality," so instead of attempting complete representation, they would select "some aspect of, or angle on, reality that would transform our vision of the whole" (p.6). This aspect of revealing the hidden dimensions of a situation or relationship was also appealing to me, as I saw it as a way to highlight our relations with the natural world which are often seen as solely as the backdrop, context, or setting for learning rather than a source for it.

Thus, I chose portraiture for two reasons: 1) the ability to document a layered narrative which includes the voices integral to the study while situating them in their social and cultural context; 2) the chance for perceptual reconfiguration, as the crafting of each portrait offers an opportunity to consider an aspect of the whole not previously considered. Knowing that the natural world in conventional notions of portraiture constitutes the "backdrop" for a human-focused portrait, I am attempting to create a form of "ecoportraiture" which builds upon questions and lessons arising from direct contact with the natural world, so that the more-than-human remains in the foreground of the research methodology itself.

Central to ecoportraiture are the following ethical starting points: that more-than-human entities each have intrinsic value and agency; that we are interdependent on the earth; that we should proceed with humility and caution. Given this, ecoportraiture is centered around several key considerations that have evolved over the course of my research; here, these are divided into two stages, one focused on data collection and the other on writing the ecoportraits. I will also describe how I developed a form of ecoportraiture, how that evolved and some of the key findings of the research process.

Developing Ecoportraiture, Stage 1: Collecting and Gathering

> Down at the river a student turns over a rock and then calls out to me, "Laura, come quick!" I find my way through the water and crouch to see what she has found. "What is it?" she asks. There is a tiny tube made of rocks that are not much larger than a grain of sand. "Look at all the colours!" the student exclaims. Each small rock is a different shade, some pink, emerald, light purple, a miniature mosaic. It's a caddisfly house, built by the larva that lives inside with its own silk and the substrate of the stream. I like the tenacity and ingenuity of this underwater architect that has carefully crafted a strong structure from what lies around it. (Piersol, 2015)

Inspired by this experience early in my research, I decided to model my portraits after the caddisfly, collecting and gathering bits of stories here and there, piecing them together to form a narrative casing to house some tender living lessons. The stories, strung in a silky web, are held up to the bright stars illuminating the caddisflies' nocturnal world,[1] revealing areas of light and dark along with room for growth and metamorphosis. With this in mind, I have separated the methodology of ecoportraiture into two stages: 1) Collecting and gathering; 2) Building the case and holding these stories up to the stars.

The aim of Stage 1: Collecting and Gathering is to make room for thoughtful choices around what to include and exclude, what is foregrounded and backgrounded, what we treat as noise versus discourse.

Collecting and Gathering with the Students

Faced with the prospect of increasing ecological crises, experts within the fields of place-based, environmental, outdoor and experiential education have spent the last 30 years attempting to instill pro-environmental behavior (Grunewald & Smith, 2008; Kahn, 2008; Warren et al., 1995). Yet, current pedagogical practices within most public schools still involve limited contact with the natural world and the programs that do exist show little evidence of long-term change in students' relations with the more-than-human world (Rickinson, 2003; Saylan & Blumstein, 2011). As previously mentioned, students are rarely in the position to be learning *from* the natural world. If this were to occur it would require re-conceiving the more-than-human world not as an inert "background" (Plumwood, 1993) but as an active co-teacher (Blenkinsop & Beeman, 2010; Leopold, 1949). To investigate this potential for the natural world to act as a "co-teacher," I chose to work with three students[2] and one teacher[3]at a public school centered on place-based, experiential and ecological theory. My central research question was: **Given the school's focus on ecological education, in what ways does student and teacher relationship with the more-than-human world change over the school year?** Together we worked to document their relations within the larger context of their learning both at home and the school.

The three students, Raven, Bambi and Ecoboy (pseudonyms chosen by the participants themselves) worked with me on a weekly basis to help collect

[1] There is evidence that caddisflies orient to the light (Grinager, 2009) and as they are nocturnal this could take the form of moonlight or starlight in calm rivers. Thus, I have used "bright stars" as a metaphor for the process of coding as it allowed me to see the separations and patterns of relation between the separate pieces of "data" in the larger narrative casing that I was crafting.

[2] I originally started with four students but one chose to drop out of the research because of personal issues.

[3] I chose to add a teacher two months into the research because I could see how the perspective of an adult educator could provide insight into the pedagogy and curriculum happening at the school.

"data" about their own perceptions and understandings as the school year unfolded. I wanted to involve the students in the research gathering stage as a way to engage them in the process, share some skills regarding the research process and potentially allow for greater freedom of expression. Each week I would engage the students in different activities (e.g., photo essays, journal entries) to aid them in collecting bits and pieces of experience that later helped me speak to their evolving relationship. At first these chats were conventional sit-down interview style but they soon evolved to be very hands-on and immersed in the place. At the beginning of the year I gave the students some brief training using some of the collecting tools (camera, video, audio recorder) and then throughout the rest of the year I would hand them over to the students to use each week, so they were actively involved in collecting and gathering along with me.

Portraiture as a methodology allows the researcher to bring together a vast array of data including participant observation (field notes, audio/video recordings, extended immersion in place), open–ended interviews of student, parents, teachers and other researchers, as well as document analysis (copies of student work, curricular documents), all of which combine to do justice to the complexity of the relationship. I started writing eco-portraits in October (see Timeline below) and then continued to do portraits of each student's relationship with the more-than-human every four months. At these intervals I then gathered "portraits" written or otherwise expressed by the students and caregivers as well, which I drew upon when writing my portraits. Every three months I also asked both the caregivers and the students to provide feedback regarding the research methodology; in the beginning stages I asked the caregivers to ask the students for feedback so that they would feel comfortable to share any concerns. After six months I asked students and caregivers for feedback directly as this seemed integral to relationship building at the heart of the methodology, and they told their parents they felt comfortable to share any worries with me by this point as trust had been built.

The main collecting tools used included: video, which captured some of the nuanced context and complexity integral to this study; audio, which can reflect tone of voice and dominant/absent voices; photography, which can speak to context; drawing, which allowed for creative expression on behalf of the students; and my own field notes/journal, which can slow the documentation down, possibly creating a more thoughtful orientation to the gathering and can travel well in inclement weather! I also collected relevant documents such as copies of student work, student drawings, and creations from home, which all helped to flesh out their relationship with the more-than-human.

My findings demonstrate that certain specific modes of data collection allowed for deeper insight into student relationships with the natural world.[4] In particular, I focused on explorations of place using context-specific arts-based methods of data collection (such as photography projects, video work, journaling, soundscapes). In addition to these place explorations, I found that seeking multiple perspectives and repeated prompts at different stages also provided essential insight into the student and more-than-human relationship. Below I will describe each of these approaches and how they were beneficial in the collecting and gathering stage.

A) Data Collection through Arts-based Place Explorations

As I got to know the students on a deeper level, I soon discovered their individual passions, interests and talents. I decided to attempt to coordinate those with the research-gathering stage as a way to engage the students in the process and possibly allow for greater freedom of expression, and that is precisely what I found. It was important that I had a range of options for investigating their relationship to the natural world as each student had different interests, and this flexibility allowed me to find modes of data collection that fit well for each individual. For example, both Raven and Bambi loved to draw and so I would focus more time on this activity with them, asking them to answer questions this way or draw together with them in free time to see what they would focus on and why. Ecoboy, on the other hand, loved photography and video so we would use these modes often to explore places together. As a way to investigate the children's relationship with the natural world, I decided to pair these arts-based methodologies with the techniques established in "go-alongs." "Go-alongs" are defined as

> a form of in-depth qualitative interview method that, as the name implies, is conducted by researchers accompanying individual informants on outings in their familiar environments, such as a neighbourhood or larger local areas. (Carpiano, 2009, p.264)

This type of interview has been shown to be useful in studying people's perceptions of place and their interactions with it (Carpiano, 2009). This approach seemed like a good fit because after the first round of sit down interviews I saw how limited the students answers were and how the natural world was still very much in the background of our conversations. As a result, I decided instead to focus on explorations of place using context-specific arts-based methods. I would often ask students to take me to a place of their

[4] I regularly checked in with the students and their families to ascertain how they were feeling about the research and changed course depending on their input. For example, if they were having a hard week for personal reasons I would ease off data collection.

choosing within the school location and then ask them to share why they had chosen that spot and what it meant to them (this has included spots by the river, deep in the rainforest, secret shortcuts, animal trails, forts, lookout towers, climbing trees). I intentionally chose place exploration as a method in order to observe the students interacting with the natural world, as I knew that we would come up against moments where they would encounter a worm or fern or log and I wanted to observe the details in terms of how they would handle it, approach it, relate with it, all of which gave me an insight into their relationship with the natural world. The explorations of place were always semi-structured where I had a list of pre-prepared questions that I would ask, often related to the place, as we explored, but I would keep the wandering open to other questions as they would arise.

Some of the arts-based approaches that I have used with the students in order to gain more information about their relationship with the natural world include:

Soundscapes. The students and I would wander around a place and I would give them an audio recorder to collect sounds that they felt represented the place. I wanted to see what they would collect and why.

Figure 4.1. Collecting Soundscapes

Memory maps. This idea was based on Hannah Hinchman's (1999) notion of an event map. The students would guide me through one of the places that they spend a lot of time in at school and share their memories of it with me. They would take drawings, photos of spots with strong memories for them and position them on a map that they have created of the place. I took audio recordings of this process. I was curious what memories would stand out and the role the natural world would play in this.

Figure 4.2. Sample Memory Map

Video work. On our place explorations I would sometimes hand over the camera to the students to let them document the place and show me around. Video was useful to capture subtleties about student interaction in the place that audio or still photos might miss.

Day in the Life Documentation. I asked the two older students to document a "day in their life" several times to see what they highlighted and recorded and why. This also allowed me to remove myself from the process for a bit to see what play-time and conversations could look like when I wasn't around. I wanted to see what a typical day looked like from their perspective and how the natural world factored into it.

Voice Thread. This software allowed me to pair still photography taken by myself or the students with audio clips from our explorations. This helped me in crafting written portraits to see the narrative.

Participant in unstructured play time. I often joined in and participated in play-time at recess and lunch and would document our adventures using

Figure 4.3. Unstructured Play Time. Note: Faces have been blurred to protect anonymity

audio or video. This would involve such activities as fort building, fire making, creating bird nests, role playing animals or climbing trees. This fostered relationship building between myself and the participants and allowed me to see their creative, playful side.

Inquiry projects. I worked with Raven to create a photo essay around her question, "Do you know what I've really, really wanted to know, if there's another colour out there? Like a new colour." She took photos here and there of colors that were new to her. In this way, I got to honor some of their own research questions, so I could position myself as a co-learner.

Figure 4.4. Sample from Student Photo Essay

Drawings. As mentioned, two of my research participants loved to draw so I would often sit with them and chat during free time as they would draw or I would draw as well. I would use drawings as an alternative to answering a question orally. This mode of expression showed me different sides of their character and created a calm, safe space for deep discussion.

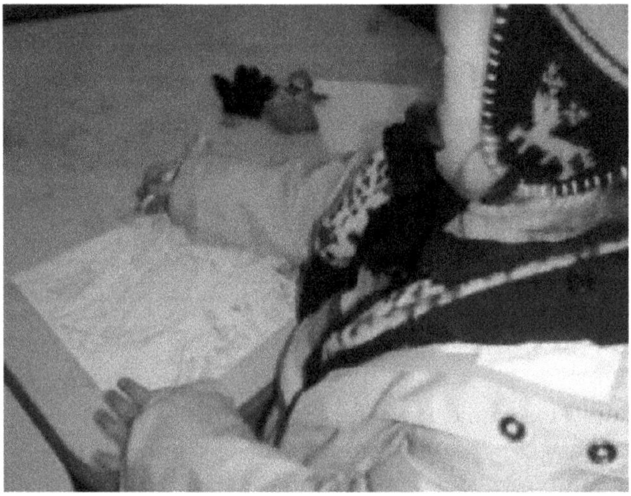

Figure 4.5. Student Drawing

Photography. I used photography like I did drawing, to help the students find other ways to express their answers to questions or show me aspects of their relationship. Often they would take "photo essays" of hikes we went on and then we would use the photos as talking points.

Figure 4.6. Sample of Student Photography

Contrast with Sit-down Interviews

When I realized that there were differences in results after these arts-based approaches I decided to intentionally ask the students the same questions that I had previously asked them during a sit-down interview in order to compare their answers and get more clarity on the differences. Here are some specific examples contrasting the sit down interview with arts-based approaches.

Table 4.1. Contrast of Sit-down and Arts-Based Interviews with Raven

Question: "How do you feel when you are drawing?"	Raven Answer
Method: Sit down interview	"Focused, like I can do a lot of details."
Method: Drawing	"Like in that picture of mine, there is a river right there. When I am drawing the river it makes me feel watery and thirsty. If there is a something missing on the drawing, I get a little feeling [tilts her head] like I've got to do that [add something]. I feel it in my body if something is missing and then I add it. When I draw a tree I feel squiggly. I feel like a tree, like if I need more branches I don't feel so good and then I just put some branches on."

Table 4.2. Contrast of Sit-down Interview and Place Exploration with Bambi

Question: "Tell me about some things you found in the forest today"	Bambi Answer
Method: Sit down interview	Silence for a long time "…bugs."
Method: Place exploration/audio recording	For full audio transcript see: Piersol, 2015

B) Data Collection through Multiple Perspective-Taking

Another form of data collection that emerged was that of multiple perspective-taking. During my research the students and I would often intentionally pose questions to each other to consider the world from the perspective of an individual creature in the natural world. For example, within the first month of my research I asked one of the students, Raven, to take me to the fort that she had been building in the woods, and while we were there I offered her a camera to take photos of it. She took about five photos of the "patio," "bedrooms" and "lookout tower" and then handed the camera back to me. I decided to try an experiment: this student had chosen the "pseudonym" of Raven because she felt connected to the birds, so I asked her to photograph the fort from a raven's perspective. Here are some notes from my field journal after this approach:

> I am struck by how she actually begins to move like a raven. She tries multiple different angles, swoops down and above. She is not concerned about documenting certain parts of the fort anymore but instead enters spaces where she thinks the raven would perch. She holds the camera for longer and takes more pictures. In my sit down interviews with Raven at the beginning of the year, she talked slowly, calmly and quietly; listening to the audio one can hear the river and a raven in the background as we are talking, but those voices remain there, in the background. In this approach I can see Raven open up and talk faster. Her eyes get wide and she begins to skip, usually ahead of me, excited to show me around. She eagerly asks me if she can use the camera. I see a whole different side of her when I allow the methodology to be more creative. We also experience a different way of being in place, where Raven, the bird, is perched on the forefront of our thoughts rather than separated in the distance.

Her photos below change from a typical human perspective on the fort…

Figure 4.7. Raven's Fort Photos from Her Perspective

to that of a raven...

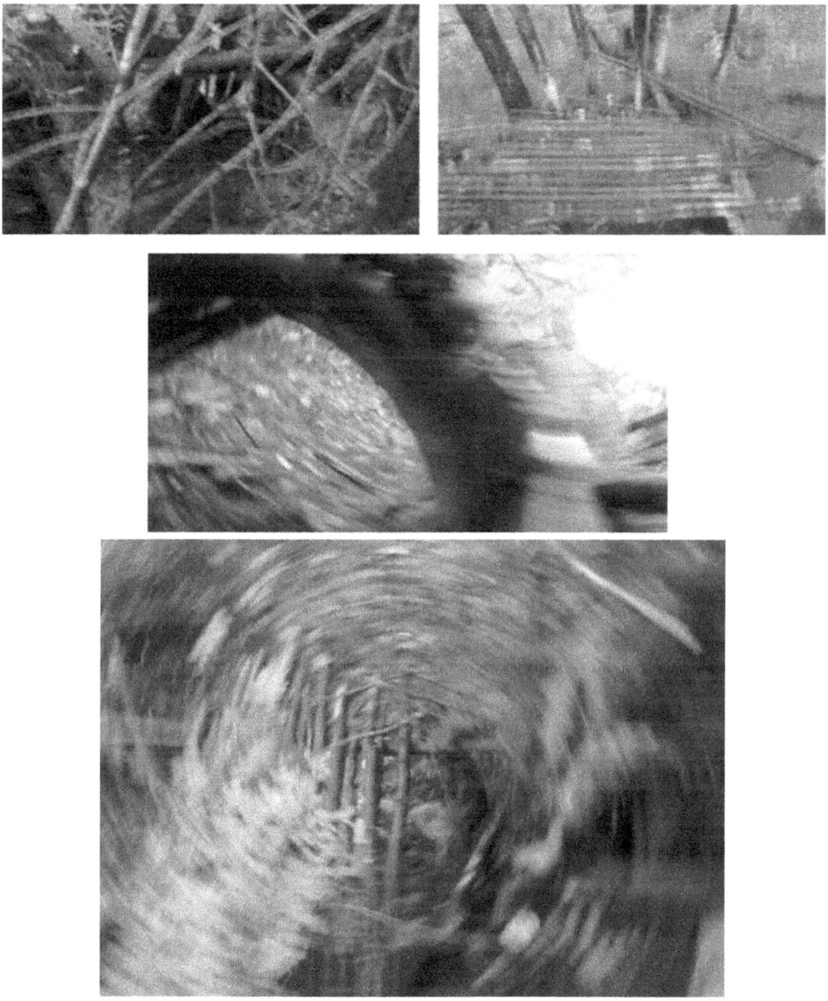

Figures 4.8 & 4.9. Raven's Fort Photos from the Bird's Perspective

In this way we were playing with learning about the bird raven's way of being in the world, what it sees and experiences, and what we can learn from it. "I like thinking of the world like it would, I know I can't understand what it actually sees but I feel like I understand it more when I try." Raven commented. Young children love to role-play, and there is evidence that this helps in empathy development as well.

Attending to voice and perspective is an important aspect of portraiture as a methodology. I wanted to include the possible perspectives of the more-than-human (as in the Raven example above) to keep it in the foreground of research, to consider what the relationships under study could possibly look like from the perspective of the raven or the slug.[5] I also purposely wanted to include multiple perspectives[6] within my portraits of the students, so that different aspects of their personalities could be revealed and highlighted, allowing the reader to see nuances and complexity. One parent comments, "You showed our qualms and that really tells me you are hearing us, and not just following your own point of view." As Michael Jackson (1989) points out, "truth and validity never reside in a single voice but always in the complex interplay of voices" (p. 191). Such attempts to highlight a multiplicity of voices add layered meaning to the relations that are integral to the research.

As previously mentioned, within this research I attempted to foreground voices that are traditionally disregarded or absent within educational research such as that of the student, parent and the more-than-human world. Instead of acting as a lone researcher, I was inspired by collaborative ethnography that emphasizes the importance of a community perspective (Lassiter, 2000; Spindler & Hammond, 2006). Lassiter defines collaborative ethnography as

> an approach to ethnography that *deliberately* and *explicitly* emphasizes collaboration at every point in the ethnographic process, without veiling it—from project conceptualization, to fieldwork, and especially through the writing process. Collaborative ethnography invites commentary from our consultants and seeks to make that commentary overtly part of the ethnographic text as it develops. In turn, this negotiation is reintegrated back into the fieldwork process itself. (2000, p. 16)

If I was to follow this method precisely, however, I should open up my research to collaboration as such right from the very start, so that parents, students, teachers and I could work to co-construct the initial questions and design of the project based on our various interests and needs. However, I intentionally did not take this approach because I want to ensure that the more-than-human world stayed within the foreground of the research. Opening the research up to try to tackle some questions and needs from the human community creates a risk that the rest of the natural world will once again slip into the background. Within educational research we become experts at creating borders and boundaries that exclude the more-than-human (Russell,

[5] I will address concerns about the anthropomorphism inherent in this process below.

[6] These perspectives include the students themselves, myself as a researcher, the teachers, the caregivers, other students, other researchers and particular entities in the natural world.

2005); this is not something I wished to perpetuate. Instead I wanted to intentionally draw attention to what Lassiter (2000) calls "the power and politics of representation," which means considering "who has the right to represent whom and for what purposes, and... whose discourse will be privileged in the ethnographic text" (p.4). I think this is a fundamental part of ecoportraiture; changing our approach to education and research to include the agency of the natural world means making time and space to listen to the voices within the natural world and consider their perspectives.

With this in mind I also tried to include multiple perspectives in the portraits I wrote. Table 4.3 provides an example.

Of course, there is concern here about the risks of anthropomorphizing when taking on the perspective of a tree or a robin, but this is where we must return to the ethical starting point of humility. In thinking about perspective-taking it's useful to consider Don Mckay's (1995) notion of "poetic attention". He describes it as follows:

> [I]t's a sort of readiness, a species of longing which is without the desire to possess, and it does not really wish to be talked about... Poetic attention is based on a recognition and a valuing of the other's wilderness; it leads to a work which is not a vestige of the other, but a translation of it... Our perception is a restructuring of the world. Poetry should not be taken to be avoiding anthropocentrism but to be enacting it thoughtfully. It performs the translation which is at the heart of being human, the simultaneous grasp and gift of home making. And the persistence of poetic attention during the act of composition is akin to the translator's attention to the original, all the while she performs upon it a delicate and dangerous transformation. (p.25–26)

I remain aware that no human language can ever speak the voice of the more-than-human, and therefore every translation has the ability to dangerously transform. We must choose our focus, our words, and our methods carefully as they help paint reality for our students and for ourselves; they help un-scale our eyes. Yes, we have a perspective as individuals and as humans, but under an ethic of care we must do our best to try to listen to the other such that we are open to how their perspective may change ours. Although we can never fully grasp the perspective of the other, an attempt is made. It is important to note that the same limitations are involved in human perspective-taking; we can never truly know what the other perceives, yet this does not and should not stop us from trying to listen. Instead of reverting to the cultural norms of ignoring or subjugating the more-than-human world within education, we need to intentionally highlight it; we need to turn our writing, teaching and research into political acts which emphasize its agency. As Annie Dillard (1974) points out, we can begin by lending an ear to these voices:

Table 4.3. Portrait Excerpt Demonstrating Multiple Perspective-Taking

Portrait Excerpt Fall 2012 Bambi
Three perspectives Flashes of light and dark are moving too quickly. It's like days and nights are moving at warp speed. Something has taken me from my dark home and I no longer know where to head. I squeeze and flinch trying to move my body to escape this place. Suddenly, I'm in serious pain, the lower half of my body goes numb. I feel much too hot and dizzy, it's hard to breathe. *—The worm* A sunny day in late winter and it's lunch time at school. I look over and see Bambi staring down at her wrist intently. I come closer and see that she has a worm curled around her arm like a bracelet; it hangs limply as she holds it by one end. I ask to see it up close and notice that part of it is squished and that it is actually drying up in the sun. It squirms under her firm grip. "Bambi, we need to put it back in its home," I say. "Look here, part of it is hurt." She drops it by accident and then quickly scoops it up and walks quickly away from me. "It needs a wet place so it can breathe properly," I follow her and tell her. "I will take care of it and protect it," she says, pinching it even more tightly between her fingers. I wince and watch as she hurries off home for the day. *—The researcher* Ever since Bambi could walk she has been playing with bugs, and cuddling with any animal she could get her hands on. Sometimes she literally cuddles them to death because she is unaware of her own strength. I'm surprised at how some of these fragile creatures are able to survive her unconstrainable love. She doesn't want poor sluggy to get left behind so she carries him for over an hour with slime oozing out of her little hands. Let me tell you, slug slime doesn't come off easy, it's just as fun to wash off as tree sap. Luckily I have a lot of experience with both. One day last spring we had had a talk about how we weren't going to bring creatures home because we didn't want to take them away from their habitat. When we got home she was really protective of her backpack and kept telling me not to look in it. Soon she got distracted and I opened it to take out her lunch and to my surprise out flies an orange butterfly. It landed on the table, we carefully put it in a bug cage and a couple hours later we set her free. It amazed me how this fragile little butterfly was able to survive being trapped in a backpack full of books moving around. How long was she in there, I wonder? Bambi will never tell. Somehow she thinks that she's going to save every animal because if she leaves a caterpillar or centipede on the ground somebody might step on it and kill it, so she's got to bring it home. But I'm trying to explain to her that if she brings all the animals home, even if it's in her little bug cage, that unless she knows what that animals eats, it's going to die at home because it doesn't have food. So I'm like, "It's better in the natural environment like where it lives, you take it away from its home it's going to be scared," and, you know... so we're still trying to go over that. I don't know if she'll ever get it. I mean look at me, I'm in my thirties and I'm still trying to bring home every animal. *—The parent*

> Two billion bacteria and many millions of protozoa and algae in a mere *teaspoonful* of soil... I might as well include these creatures in this moment as best I can. My ignoring them won't strip them of their reality, and admitting them, one by one, into my consciousness might heighten mine... (p. 96).

Explorations in place also helped to facilitate a sense of multiple perspectives and led to more inclusive conceptions of voice. This can be seen in the transcript of lunch time exploration of place with Bambi. The worm, the mosquito, the robin all entered into the foreground of our consideration, which is much different from our experiences just sitting with each other talking. As these "voices" and perspectives of the worm or the raven entered the data collection stage through place exploration, I got to see how the students approached them, how they listened or attended, what they noticed or ignored, and this in turn gave me deeper insight into their relationship with the natural world.

C) Data Collection through Repeated Prompts

In addition to seeking multiple perspectives and engaging in explorations of place using arts-based methodologies for data collection, I also designed some questions to help me assess change over time within the student/more-than-human relationships. This helped me to address my main research question, "In what ways does the student and teacher relationship with the more-than-human world change over the school year?" Every four months, with caregivers, teachers and students, I would ask some repeated questions such as, "Describe the student for someone who has never met them before," or "Describe some examples of interactions with them in outdoor settings." I intentionally chose open-ended questions that would provide a sense of where the student was at in that moment in time in terms of their relationship with the natural world. These questions acted as guideposts to help me see change over time; I kept track of individual answers and noted how the answers changed as the months progressed. I followed the same approach with students, asking them every 4 months about their relationships with the natural world using the same questions every time. This was very revealing and this process helped to turn portraiture from a static snapshot to a more fluid sense of time.

Developing Ecoportraiture, Stage 2: Building the Case and Holding the Stories Up to the Stars

The aim of Stage 2: Building the case and holding the stories up to the stars is to make room for ongoing reflection on methodology and eco-portrait creation and then adjust things accordingly based on feedback.

I had originally envisioned revising the methodology only after gathering participant feedback and input, yet I found that I was constantly tweaking my approach week to week given my experiences working with the students. Every four months I sent the portraits that I had written out to the students and caregivers to provide feedback and edits;[7] I would then revise where needed and send them back out to participants for another look. I also created a list of questions to assess the students' changes in relationship over time that I thought would be useful in keeping the natural world in the foreground of this reflection stage, such as, "Are we returning to different times of year, times of day?", "What voices are absent/ignored?", "Are we taking time to write details, particularities, slow pace?" However, as I mentioned above, by shifting my methodology to include context-specific arts-based approaches of exploring place, the consideration of multiple perspectives, and ways to track changes over time, I found that those questions were not needed as this change in methodology was addressing them. It became apparent to me that regularly adjusting and adapting the methodology in response to context was essential in keeping it rooted, updated and related to all of those involved in it.

After collecting and gathering bits of stories here and there with the students in the above manner, it was now time to stick them together to form a narrative casing to house some tender living lessons. I will describe this process below and some of the bright stars that I held the stories up to, revealing room for growth. I wrote multiple portraits over the course of a year, as these stages were cyclical. After I determined where gaps could lie, I began to gather more bits of stories to add to the casing. Eventually I had so many bits of story tied on that the narrative casing seemed strong enough to be a fit house for the lived experience I wanted to share.

Sorting the Stories Gathered

Portraiture suggests five different processes in the crafting of themes: 1) listening for repetitive audible and visible refrains (repeated behaviors, words or statements); 2) seeking resonant metaphors which act as symbols for larger stories happening in the lived experience of the participant; 3) discovering different cultural, individual and institutional rituals which may shed light on values and narratives of participants; 4) triangulating the data from a variety of sources; 5) attending to dissonant voices which may contradict themes or

[7] There were a few instances where I tried to be mindful of personal issues families were facing, as I did not want to bombard them with information during this period. Thus, in a few cases, this timeline was not met.

patterns and in this way flush out complexity and particularity of the narrative. I have re-named these five processes "bright stars" as they were key in guiding and illuminating the stories that I had collected. Table 4.4 is a sample of an interview under the light of some of these bright stars.

From this transcript we can see examples of repeated refrains such as "caring" and "wanting to share" for Ecoboy. These are examples of words or statements that I as a portraitist heard many times over, from various people, often including the research participants themselves. Through this example one can also see some of the resonant metaphors, for instance "making connections" and "calming." Lawrence-Lightfoot and Davis (1997) point out "these words or phrases resonate with meaning and symbolism, sometimes representing the central core of an institutional culture or the dimension of a life story" (p.198). Such metaphors are often poetic and symbolic expressions of the lived experience of the participant. In this case, "making connections" seems to be a driving metaphor for Ecoboy in terms of how he approaches life and learning, but it also speaks to his ability to see the interconnected nature of things. The same is true for "calming" which seems to be symbolic of his life's path at that moment in time. This interview also provides an example of a "deviant voice," thoughts or perspectives that seem to contradict past statements.

> As the researcher organizes transcript data into categories, she must be vigilant in attending to the experiences and perspectives that do not fit the convergent patterns. In portraiture, we refer to this perspective that deviates from the norm as 'the deviant voice' and we never stop listening for it, even as we become increasingly focused in our inquiry and certain in our analysis (Lawrence Lightfoot and Davis, 1997, p.193).

These are not always explicit; the researcher often has to sift through with a critical eye to what is not said or is absent in observations or discussions to find this voice. In this case, a teacher who regularly expressed positive views of student behavior at the school noted that "some kids were just walking up and destroying stuff." Seeking this deviant voice[8] is useful in helping me to see the other side of things and in doing so helps me to seek new ways of framing or understanding the lived experience of the research participant. I also labeled transcripts for "rituals," events that could reveal values, priorities and stories important to the participant. This type of sorting using the

[8] Lawrence Lightfoot and Davies (1997) describe this deviant voice as "experiences or perspectives that do not fit the convergent patterns" (p.192–3). In this instance, the teacher's disparaging comments were atypical from the pattern of highly positive remarks that I usually received from him.

Table 4.4. Teacher Interview as Seen under the Bright Stars

Interview with Teacher Transcript Sample	Ecoboy Fall 2012 – Bright Stars*
Observations re: natural world interactions	
Teacher: Yesterday we were talking about "Whose home is this?" because some of the kids were walking up and just destroying stuff. He answered very quickly when I started asking about "what's that log?" He quickly had two or three answers that he wanted to share and recognized that it was more than just a rotting log.	Some kids "destroying stuff" (deviant voice) "Wanted to share" (*repeated refrain*) "Recognized it was more than just a rotting log" (*resonant metaphor:* making connections)
Me: Any changes that you've noted in Ecoboy over the last school year? Examples?	
	"More comfortable"/"confident" (*resonant metaphor*; calming) "Not very confident at all" re: schoolwork in the past (repeated refrain)
Teacher: He feels more comfortable and therefore confident, in the past he's not been a very confident child when it comes to schoolwork. He wants us to scribe, his ideas are good and I think that he could do more but I think that he lacks that confidence but I think that it will come. He doesn't go into very much depth in anything written, so you know orally he'll do more, it would be a good way to test him. I think he's very caring. Yesterday C was being mean to him; he didn't bring that up necessarily but dealt with it well. Ecoboy's a good influence; he's always kind of calming. I see him being this way with friends but it's also how I see him in general right now, like he is calming himself, becoming more comfortable, less anxious.	"Caring" (repeated refrain) "Calming" (*resonant metaphor*: calming)

*Note: Bright stars used for codes indicated in parentheses.

bright stars was done for each season of the research term. All data was coded in this manner.

Crafting the Themes

Using data from field observations, parent/teacher/student interviews, and student work, triangulation was used to craft emergent themes through coding for the bright stars of visible and audible refrains, resonant metaphors, dissonant threads and embedded rituals (Lawrence Lightfoot & Davis, 1997). Similar codes found under these bright stars were grouped together and from this, themes were discerned. Table 4.5 is an example of an emergent theme crafted from the coding process.

Table 4.5. Coded Terms Sorted into Themes

Ecoboy Themes Fall 2012	Terms coded using the Bright Stars
Caring for human others	Calming (resonant metaphor) Making Connections (resonant metaphor) Smiling, nodding (visible refrains) Patient (audible refrain) Encouraging (audible refrain) "Wants you to be happy" (*audible refrain*) Praise from teacher (*audible refrain*) Helping (*ritual*) Sweet (*audible refrain*) Sensitive to others (*audible refain*) "Natural leader/helper" (*visible and audible refrain*) "Wants to please" (*dissonant voice*) Not listening (*deviant voice*) "Goofing off" (*deviant voice*)

Here we see multiple codes grouped together due to similarity. From this emerged the title "Caring for human others" which acted as an underlying theme for all of these codes. The themes were then used to help structure the portraits that I wrote every four months (see below for example). I found grouping codes for themes to be extremely helpful and although I was worried about the possibility for it to simplify things into discrete categories, I found it instead allowed me to observe connections and realize how complex the relationships actually are. Thus, I identify with this point made by Lawrence-Lightfoot and Davis (1997):

> Naming convergence within a complex and diversified whole may seem to be a reductive activity. But the identification of emergent themes does not reduce the complexity of the whole; it merely makes complexity more comprehensible. (p. 215)

In addition to the coding that I did in this manner, I also did some interpretive reading (Gilligan, Brown, & Rogers, 1989), consisting of multiple readings of the same transcript, listening for a different voice[9] or angle each time. This helped me to look for connections among themes, subtle changes over time, and try to maintain the particulars and the complexity of the relationship rather than flattening it into broad categories (Miles & Huberman, 1994). I used the themes that I constructed as a frame for the writing and then tried to flush out the details, particulars, subtle changes and dissonant voices through the narrative. In this way, I tried to balance the tension that those using the methodology of portraiture must face:

> ...the tension between organization and classification on one hand and maintaining the rich complexity of human experience on the other—the tension between developing discrete codes and searching for meaning, and the tension between the researcher's desire for control and coherence and the actors' reality of incoherence and instability. The portraitist does not try to resolve this tension by choosing one side over the other. Rather she works to maintain the tension and experience the dialectic between these two approaches to thematic development (Lawrence Lightfoot & Davies, 1997, p.192).

Table 4.6 provides an example of how I attempted to deal with such tension as I wrote the portraits. The theme that was structuring this piece was "A companion in wonder," referring to a support teacher at the school who regularly encourages and takes interest in Bambi when she finds items in the natural world. Instead of solely describing how Rosie acted to facilitate wonder, I wanted to portray some of the tension that Rosie felt, trying to get Bambi to keep up with the group and yet still trying to honor her finds.

Although coding using the bright stars helped me to discern the theme of Rosie's role in encouraging Bambi's wonder, I also chose to include subtle details in the narrative showing how Rosie constantly tried to get Bambi to catch up and focus back on the teacher. These subtle encouragements did not often show up on the audio as they were a wave of the hand or a soft turn of the body to face the teacher, so they could've been easily missed had I just relied on coding the audio alone. Instead, I remembered subtle details about the walk that I included to balance the extreme patience that Rosie embodied with the impatience that she mentioned she felt inside, as a way to illustrate

[9] These voices included the more-than-human, the parent, the child and the teacher.

Table 4.6. Excerpt from Portrait of Bambi

Excerpt of Portrait under theme "A Companion in Wonder" – Bambi Fall 2012
A crisp, sunny fall morning and today the class cuts onto a trail in the woods. Before we enter I help one of the older boys take a picture of a spider he found on one side of the road. Bambi remains behind with him as the rest of the students enter the forest. Rosie stands by the cut-off and waves for Bambi to speed up and catch the group. I go over to help encourage Bambi to catch up and overhear her ask the older boy, "When you are done taking that picture show it to Sam, OK?" She knows that her friend Sam, also in grade one, loves bugs and so she wants to be sure that someone lets him know about the spider. I tell this to Rosie as we wait for them to catch up. Rosie says, "See! You just have to listen to them." I remark that yes, Bambi wasn't simply lagging behind but rather was being quite thoughtful on behalf of a friend of hers. "I don't think I'm patient enough," Rosie says. This surprises me because I consider her to be an extremely patient person. The rest of the hike Rosie walks with Bambi and another kindergarten student. I hear Bambi soon yell out, "Rosie! A slug!" "Wow," says Rosie, crouching down to peer at the creature with her. Rosie looks up the trail; realizing that they are losing track of the rest of the group, she urges Bambi to keep moving. Bambi says, "But I want to move it out of the way." Rosie reassures her that the slug will be safe there. Twenty seconds later, Bambi shouts out again, "I found something. Look! Look!" I turn back to see she has found a hole in the ground the size of a golf ball. "I know what made that," Rosie says and looks closely with her. We ask Bambi what she thinks could have made it. "A mole," she says. Bambi stops a few steps later to touch some shelves of fungus. "Fairies might live here," she says. Rosie greets each of Bambi's discoveries with tenderness, care and attentiveness. I notice that Bambi, as she makes her way along the trail to catch up, points out slippery spots or sticks to Rosie, saying, "Rosie, it's slippery here, hold hands," or "Rosie, careful, this is a hill." We make our way to the bottom of the hill and finally meet up with the class; Bambi has found plastic ties and tells me they look like "rabbit ears". I would have never thought of that. Ocean begins talking to the class and off to the side I see Rosie bend down to look at a piece of wood that Bambi is showing her. "How lovely!" Rosie says and they grin at each other. After this, Rosie gently guides her attention back to the teacher.

some of the complexity of the situation. Through the narrative approach of portraiture, I felt able to convey dissonant voices and rich nuances so that as these themes emerged, they remained filled with complexity that at times defied their own discrete label.

Using Themes as Springboards Forward

As I did ongoing coding, synthesizing, reflecting and writing throughout the data collection stage, this work acted like a springboard pushing me forward, revealing gaps that I could refine and identifying key areas to study next when collecting and gathering. Miles and Huberman (1994) called this ongoing work with coding "iterative cycles" that assist with clarity and refining themes. For example, Table 4.7 shows some questions generated after writing a portrait about Bambi under the theme of "Attentiveness." I also identified some possible sources to help me answer these questions. This process really helped to refine my inquiry, and I would then use the results of this stage of the exploration to create the next set of questions for parents, teachers, field observations etc.

This process helped to provide refinement and clear paths forward for each stage of data collection. That being said, many themes were abandoned for a variety of reasons; some were not as significant or strong as others, some were off topic from the general research focus of relationship with the natural world, and some simply stopped appearing. As a result of starting with broad questions, I was able to craft many themes initially in order to guide collecting and gathering, and was never at a loss for questions or threads to pursue. This is in line with one of the goals of portraiture:

Table 4.7. Questions Arising out of Research & Possible Sources for Answers

Bambi "Attentiveness" Theme Fall 2012 Questions	Possible Source
What is it about the early childhood years that lends itself to that wide-eyed awareness?	Literature, question for other researchers, parents
Is this attentiveness innate? Are/were her parents like this?	Question for parents, literature
Is her ability for keen observation a skill that has been cultivated? By whom and how?	Question for parents, teachers daily observations of those that might assist with this at home (read through parent journal) and at school
How does she retain this skill when most adults around her are encouraging her to rush her pace and focus in on human interaction?	Field observations and question for teachers, parents
What does this "detailed eye" extend to?	Observe her drawings, what she halts for and why, question for teachers and parents

> [A]ny system of data organization and synthesis must be flexible enough to allow the researcher to shift gears and change direction as she moves from fieldwork to analysis and back to data collection. (Lawrence Lightfoot & Davis, 1997, p.188)

Story-based Representations as Complex and Ethical Projects

To demonstrate what an "eco-portrait" might entail I will share in Table 4.8 an excerpt of a portrait of Ecoboy crafted around the theme of "attentiveness" that appeared for him. Following this, I will share how the final stage of "case building" for ecoportraiture is intentionally a complex and ethical endeavor.

Table 4.8. Excerpt from Portrait of Eco Boy

Excerpt from portrait of Ecoboy, Fall 2012
We walk toward the fort village and Ecoboy stops at a muddy patch to look at the tracks. "Looks like dog," he says. I agree and we look for claw marks. Satisfied with our identification, we keep walking. He is walking in front of me and again I'm surprised when he goes off trail away from the fort village. He detours to take a look at the river. "That would be good cover for a fish," he says, pointing to the overhanging tree branches. "There's a dead fish below, you can see it there?" he asks. "Oh neat," I say, looking to see the body caught in a whirlpool created by a mossy log. As we are leaving Ecoboy points a large hole in a log below our feet; "Maybe that's a home for a marten," he decides. As we enter the fort village he stops and looks in another hole at the base of a tree. I'm surprised by this because he's walked this route so many times, and yet he still stops to look closely at things as he is walking. He reminds me of his fort from last year and we decide to go check on it. I follow behind him. "I haven't been back here in a while," he says. "There's the bird's nest." He points out a mossy ball hanging off a tree branch. "I think it's abandoned now," he says, peering up at it. He takes a photo of it. "If I had my waders on I'd go down in there." He looks at the pool of water where his old fort used to be. I laugh and say, "Be careful," as I watch him cross a slippery muddy slope. "There's probably frogs in here now," he says. We look into the pool, checking for that splash of green and listening quietly to see if we hear any movement in the water. As we wait, I realize that instead of simply showing me his fort, Ecoboy has been stopping along the way to show me a community of living creatures that also share this spot and call it home. The tracks in the mud, those fish in the river beside the village, the bears that he has seen pass through, those holes at the base of the trees, that mushroom growing on the fort, the bird's nest, the frog pond, he has taken the time to pause and attend to all of these things. I normally walk quickly on the trail all the way here and he asks me to divert off of it. I normally only see the human impact on the area in the fort village and he asks me to crouch down and look at the small mushroom I would've easily missed. I thought we were going on a walk to hear about his fort when really Ecoboy had been busy soaking up the signs and clues that surround him.

I liked playing with the aesthetic dimensions of story because, as previously mentioned, crafting a portrait can help shift our distribution of the sensible: to reveal aspects of the whole not previously considered. Hopefully the result is something that inspires the reader to look at the world in a different way. For example, in the above portrait with Ecoboy I tried to highlight Ecoboy's way of relating to the world around him such that the reader could see the forest from his perspective and hopefully, in doing so, experience a metaphorical shift from seeing the forest as "background" to seeing it as a varied and diverse place, full of life, full of homes and worth stopping to explore.

How the narrative is crafted depends on many factors. Lightfoot and Davis make a distinction about switching from "listening to" the story towards "listening for" the story (1997, p.99). In the traditional methodology of portraiture, the researcher is listening *for* a story, actively creating a narrative that is both authentic and open to interpretation. They are careful to note that this is *a* story, not the "*the*" story. However, I was wary about this switch, as I'm constantly asking the question, "Is our purpose here getting in the way of what place or the more-than-human world has to teach/tell?" I was worried that if I began "listening for" a story, if I had my own agenda about what the story should be, then I wouldn't be able to really listen closely to how the natural world was actively influencing our ways of knowing and being. As a result, as I explored place with the students, I decided not to listen "for a story" per se but I rather "listened to" the students and the place quite intently. However, interestingly, afterwards as I listened to the audio or read the transcripts I switched to the "listening for."

> Post-data collection analysis is less action-oriented and more ruminative than the day to day analysis of the data gathering phase—less about preparing for the next day of purposeful and strategic data collection and more about deep contemplation and probing insight (Lawrence Lightfoot & Davis, 1997, p. 189).

By holding the bits and pieces of information that I'd gathered up to the bright stars that portraiture offers, I was able to step back and see the larger picture and hear how the voices intertwined, I could see patterns in the tales collected, and out of this I begin to see larger themes. For example, in the above portrait about Ecoboy, I originally went into the interview thinking it was going to be about his fort and filed it under that heading in my files. However, as I listened to the audio, glanced at the photos and read my notes, I soon realized the story that was unfolding was more about his relationship with the natural world.

By stepping back from the need to construct a story from the outset and listening rather to what emerged, I felt that I'd been learning how to co-create the story *with* place and the students. Often a student interrupted my "plan" or questions; for instance, one day I was trying to chat quietly with Ecoboy and he started throwing pebbles to see if he could reach the river from where we were sitting. Instead of just chatting we broke out into laughter as we challenged each other to throw them farther. He changed the narrative arc of that day and I was grateful for it. The same is true for the natural world; when I was with Bambi, the grade one student that I worked with, inevitably a slug or the worm pushed its way by and changed where the story with her was headed and what emerged. This co-creation of story allowed for various voices to be highlighted other than my own, and the flexibility of letting the students take control of where the story should head helped build relations between us. In this regard they feel like co-researchers rather than studied "subjects".

A storied approach also meant that I tried to note as many details as possible throughout the day, from the shoes of the student to the color on the wing of the bug, because I never knew what was going to be important or not. Of course I couldn't pick everything up, but portraiture asks us to seek out relations in a detailed way that I don't think I would have captured if I was simply making general observations. In this methodology, the challenge ends up being that of finding out how the rock throwing is connected to all the rest. For example, as I crafted the story above, I grew curious as to how little moments, like Ecoboy waiting quietly in anticipation of finding a frog or gently touching a mushroom, could possibly be connected to larger themes of empathy and curiosity that have arisen for him. It helped me to think in terms of context, the body, relation and the particular, all key elements of what I came to think of as eco-care.

I also found that the process of portraiture revealed as much about the researcher as it did about the participants. It suggests that we all may be brought, through the research process, to see aspects of ourselves that we may not be used to having out in the "light of day". As I shared at the end of the portrait about Ecoboy, here I am an "ecological educator" of 18 years, and yet I had been walking that trail that I walked with Ecoboy countless times, not noticing even remotely the extent of things that he picked up in a couple of minutes. In trying to *know* the "other," I came to recognize depths of being and knowing that I could never quite access. I found the process humbling, as it put me steadily in the position of "student" or "learner" as much as "researcher." Portraiture helped me to see research in a new light, as an endeavor that is inherently messy and paradoxical but also rich because of these very complexities. One of the parents I work with hinted at this very thing when she described her appreciation for "the way you see all the angles, and the way you let it all be both oppositional yet pull it all together." It means that when writing or interpreting or translating, there will always be another way to look at it, another angle or perspective to consider.

Conclusion

The lessons around ecoportraiture articulated here are contingent upon one another and not meant to stand alone. For example, incorporating multiple perspectives and voices could simply result in moral relativism if not paired with certain ethical starting points such as acknowledging the agency of the more-than-human world. A "complex understanding" void of context could result in daunting abstractions leading to apathy. I want to be clear that these examples are by no means a comprehensive "list" of ways into ecological portraiture and that they are not meant to be taken as a "recipe." Assuming that one can simply outline the characteristics of ecological consciousness and then embody that in a methodology simply replicates the de-contextualized, simplified models for thinking that are at the root of the ecological crisis. Instead, ecoportraiture emphasizes the need for contextualized, diverse and complex understandings that take into account the local and the particular instead of the generalized and abstracted. As a result, the lessons offered here are not meant to act as a definitive prescription but rather, as starting points to critically consider the current distribution of the sensible regarding the natural world within schooling and research. How does methodology shapeshift if we "become aware" of the more-than-human world as having agency, and thus as also having the ability to inform and teach? My hope is that ecoportraiture can help us re-imagine what is speakable, thinkable or doable in schools and in research, and finally make room for silenced voices of the more-than-human to be heard and considered.

References

Blenkinsop, S., & Beeman, C. (2010). The world as co-teacher: Learning to work with a peerless colleague. *The Trumpeter: Journal of Ecosophy*, *26*(3), 27–39.

Carpiano, R. M. (2009). Come take a walk with me: The "go-along" interview as a novel method for studying the implications of place for health and well-being. *Health & Place*, *15*(1), 263–272.

Dillard, A. (1974). *Pilgrim at Tinker Creek*. Toronto, ON: Bantam Books Inc.

Gilligan, C., Brown, L., & Rogers, A. (1989). Psyche embedded: A place for body, relationships and culture in personality theory. In A. Rubin, R. Zucker, R. Emmons, & S. Frank (Eds.), *Studying persons and lives* (pp. 86–147). New York, NY: Springer.

Grinager, J. (2009). *Caddisfly larvae visual system: Reponse to light.* Madison, WI: University of Wisconsin.

Grunewald, D., & Smith, G. (Eds.). (2008). *Place-based education in the global age: Local diversity.* New York, NY: Lawrence Erlbaum.

Hinchman, H. (1997). *A trail through leaves: The journal as a path to place.* New York, NY: W.W. Norton & Company.

Jackson, M. (1989). Paths toward a clearing: Radical empiricism and ethnographic inquiry. Bloomington, IN: University of Indiana Press.

Kahn, R. (2008). Critical pedagogy, ecoliteracy, and planetary crisis: The ecopedagogy movement. New York, NY: Peter Lang.

Lassiter, L. (2000). *The Chicago guide to collaborative ethnography.* Chicago, IL: University of Chicago Press.

Lawrence-Lightfoot, S. (2005). Reflections on portraiture: A dialogue between art and science. *Qualitative Inquiry, 11*(1), 3–15.

Lawrence-Lightfoot, S., & Davis, J. H. (1997). *The art and science of portraiture.* San Francisco, CA: John Wiley and Sons Inc.

Leopold, A. (1949). *A Sand County almanac.* Toronto, ON: Random House Inc.

McKay, D. (1995). Baler twine: Thoughts on ravens, home and nature poetry. In T. Lilburn (Ed.), *Poetry and knowing.* Kingston, ON: Quarry Press Inc.

Miles, M., & Huberman, A. (1994). *Qualitative data analysis: An expanded sourcebook* (2nd ed.). Thousand Oaks, CA: Sage.

Piersol, L. (2015). *Eco care: Nurturing possibility and resistance in education.* Unpublished doctoral dissertation, SFU.

Plumwood, V. (1993). *Feminism and the mastery of nature.* New York, NY: Routledge.

Rickinson, M. (2003). Reviewing research evidence in environmental education: Some methodological reflections and challenges. *Environmental Education Research, 9*(1), 257–271.

Russell, C. (2005). 'Whoever does not write is written': The role of 'nature' in post-post approaches to environmental education research. *Environmental Education Research, 11*(4), 433–443.

Saylan, C., & Blumstein, D. (2011). *The failure of environmental education (and how we can fix it).* Berkeley, CA: University of California Press.

Spindler, G., & Hammond, L. (2006). *Innovations in educational ethnography: Theory, methods and results.* Mahwah, NJ: Lawrence Earlbaum Associates.

Warren, et al. (1995). *Theory of experiential education: A collection of articles addressing the historical, social and psychological foundation of experiential education* (3rd ed.). Dubuque, IA: Kendall Hunt Publishing.

5. *Indigenous Knowledge as Voice: Ecoportraiture and Indigenous Ways of Knowing*

Sylvie de Grandpré

> "The works must be conceived with fire in the soul but executed with clinical coolness."
>
> — Joan Miro

Introduction

Sara Lawrence-Lightfoot and Jessica Hoffman Davis (1997) introduce the notion of *essence* in the opening pages of their book *The Art And Science of Portraiture.* In particular, Lawrence-Lightfoot describes her understanding of the artist's intention to paint beyond what is simply given, bringing into the light attributes of a more abstract nature while constructing her portrait (p. 4). She names this motivation as a desire to capture the essence that emanates both from her as artist but also from the developing relationship between the artist and her subject (p. 4). The portrait is not simply a painted photograph, a record of the visible, but rather a rendition made richer by emphasizing the relationship between both the physically noticeable and the impalpable subtleties: the feel, the vibe. The authors draw attention to the interactions between elements of expressive content, pointing to the importance paid to "each aesthetic aspect as well as to the relationships among them" (p. 29). *Essence*, therefore, implies attending to each element of a subject and to the relationships between each and all of these elements.

This chapter grew out of my efforts to apply my understanding of *essence*, originally developed in the context of my own culture, to the challenge of embedding Indigenous ways of learning into the curriculum as a non-Indigenous person. Through most of my thirty year career I have taught

French immersion or French as a second language in British Columbia (BC). It's not just about the food, songs, or traditions, it's about the myriad of subtleties that contribute to a better understanding of the complex cultural background while learning to speak French. To communicate these subtleties can be challenging for any teacher. However, with enough contextual exposure and experiences, students and teachers alike can begin to understand the "how" or the "why" of a particular culture.

Working to embed Indigenous ways of learning into my teaching has demanded reflection and growth from me as a teacher and as a person. Indeed, as an ongoing process it continues to call for more purposeful, conscious, and in-depth consideration than my habitual sharing of my own culture. In part this is because my understanding keeps shifting along with my practice. As Nadine Johnston notes in McGregor, Restoule and Johnston (2018), "I come to this work as a learner and as a helper" (p. 3).

Since the beginning of my career, I have included Indigenous *content* within my teaching, initially through the BC Social Studies curriculum. With the creation of our school district's first Aboriginal Education Enhancement Agreement (AEEA) in 2005, I began including Indigenous content while teaching about French culture by working along with the Aboriginal Education Support Worker (AESW). In the last eight years, I have transferred out of French immersion and started teaching mainstream English. This new teaching endeavor has coincided with the implementation of a new provincial curriculum and the inclusion of the First Peoples Principles of Learning (FPPL) throughout the curriculum. I accepted this as a call to develop a heightened awareness of Indigenous ways of learning and a deeper understanding of Indigenous worldviews.

Here, I must acknowledge that my learning of Indigenous ways began to develop upon moving to BC, living on Ktunaxa territory, and working as a teacher, through many experiences. Early in my career, I had the opportunity to meet and work with Nasuʔkin Joe Pierre Jr. (Chief Joe Pierre), who has been an Aboriginal Education Coordinator for our school district since the creation of the first Enhancement Agreement. More importantly, he is a captivating storyteller. For the past twenty five years our career paths have been essentially parallel, but whenever we have re-connected our relationship has remained positive. Prior to starting my PhD, I sought Nasuʔkin Joe Pierre's advice about including Indigenous students within my studies. He counseled a *prise de conscience* of assumptions and colonialism. Later, in 2018, as I requested his participation on my committee, he shared a story and a metaphor. The story was about never stopping to learn, the metaphor was about the FPPL. At the time, I expressed my interest in looking at how teachers in our district, predominantly non-Indigenous teachers, can embed Indigenous ways of learning more meaningfully. He responded that a thin blanket was laid over BC when the FPPL were included in the curriculum and that such work would contribute to making it a little thicker.

In a previous research project (de Grandpré, 2017), I looked at the subtleties involved in embedding Indigenous ways throughout the curriculum, more specifically the FPPL (Figure 5.1). The principles are guiding learning statements, developed under the auspices of the First Nations Education Steering Committee (FNESC), that represent BC Indigenous ways of learning. Through the nine principles there is a sense of culture and place specificity that encourages relevance at the local level. For example, one of the principles indicates that some knowledge is sacred unless "shared with permission and/or in certain situations" (FNESC, 2008). Consequently, some local knowledge requires pre-approval, either from the AESW or the district's Aboriginal Education Coordinator, and ultimately the relevant First Nation community, before incorporation into a lesson. Embedding the principles requires purposeful thinking and planning rather than merely matching principles to lessons or activities. Embedding the principles requires that teachers authentically engage with Indigenous ways of learning and/or knowing.

The results of the FPPL project highlighted the importance of being and learning "in place," and helped identify an array of interrelated factors involved in embedding Indigenous ways of learning into the curriculum. Eight themes or categories emerged (perceptions, connecting, understandings, communication, strategies, challenges, awareness, and holistic), each of which was closely related to the next while holding some specificity of its own. Often, elements from one category affected elements in another or worked in conjunction with elements from several other categories simultaneously. Together, the categories provided a glimpse of the essence within Indigenous ways of learning. As Wilson (2008) remarks, "one of the problem is that the elements of an Indigenous paradigm are interrelated or interdependent; it is difficult to separate one to write about" (p. 69). He further explains that for everything that is, there exists a corresponding web, or "knot," of relationships, and specifies that "some of these knots of relationships are not visible or tangible entities" (pp. 76–77). For Wilson, "this is our epistemology" (p. 77). Developing the ability to start recognizing and begin to understand those knots of relationships is key to embedding Indigenous ways of learning in one's teaching practice; it is also the promise that portraiture methodology holds out for me through its emphasis on essence.

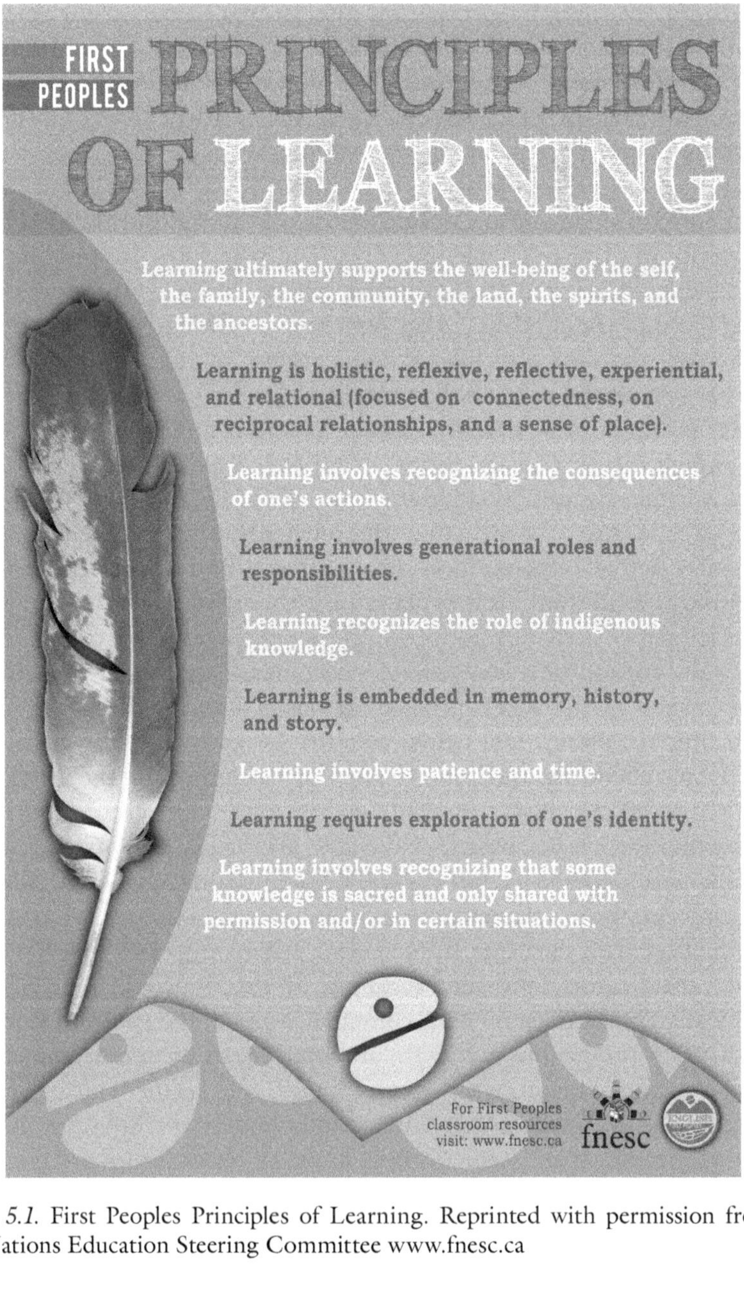

Figure 5.1. First Peoples Principles of Learning. Reprinted with permission from the First Nations Education Steering Committee www.fnesc.ca

Throughout this chapter, my use of the term "Indigenous research" is informed by the perspective developed in Wilson (2008). My thinking has also been influenced by recent publications discussing Indigenous research and the clarification that Indigenous research, while representing a wide variety of Indigenous paradigms, is differentiated from non-Indigenous research (McGregor et al., 2018). I use the term "Indigenous ways of learning" to refer to the FPPL, the development of Indigenous knowledge, and to the manner in which knowledge is acquired through multiple perspectives (Wilson, 2008). The intent is to also encompass the notion of fluidity expressed in the term "coming to know" (Cajete, 2000), as discussed by Belczewski (2009) and Sutherland and Henning (2009). At times, the term "Indigenous knowledge" is more accurate, thus it is used instead.

Relationality, relationships and relational accountability are terms that hold paramount importance within Indigenous research and encompass the epistemology, ontology, axiology and methodology of Indigenous research *together*, never apart from each other (Chilisa, 2012; Wilson, 2008). In relation to Indigenous research in particular, Wilson (2008) summarizes it as involving "thinking of the world as a web of connections and relationships. Nothing could be without being in a relationship, without its context. Our systems of knowledge are built by, and around and also form these relationships" (p. 77). This brings into view the circular manner in which Indigenous research functions. It is not a "methodology" in the sense that portraiture is a methodology, but a fluid and deeply connected process of relational knowing (see below).

Finally, in referring to anyone partaking in research I will use the words actors or participants, sometimes interchangeably, as they are referred to by various authors. Ultimately, all these people(s) are co-producers of knowledge.

Connecting Portraiture with Indigenous Research

The reality of bringing a Western methodology, such as portraiture, and Indigenous research together is that only parts of the whole, not the entire whole, can become more deeply understood by the other paradigm (Belczewski, 2009). Researchers further note the difference between the linear/hierarchical model of Western research methodologies and knowledge, and the circular functioning of Indigenous research and Indigenous ways of learning (Hatcher et al., 2009; McGregor et al., 2018; Styres, 2011; Styres & Zinga, 2013; Wilson, 2008). As of yet there is very little literature connecting portraiture to Indigenous research, so it is important in any comparison to keep in mind key questions, such as who owns the research process, what are the unique characteristics and essences of Indigenous research, and what tensions and nuances with portraiture need acknowledging (Chilisa, 2012; Kovach, 2018; McGregor et al., 2018; Smith, 2012; Wilson, 2008). The key is not to see if one can merge with the other, but rather to see if there are enough points of connection for mutual enrichment.

Indigenous research is indissociable from Indigenous knowledge. Indigenous knowledge is locally dynamic in that it continuously evolves, is thousands of years old, and has been transmitted orally, through culture, and through lived experiences, for generations (Chilisa, 2012; McGregor et al., 2018; Wilson, 2008). Further, and as significant, Indigenous knowledge is circular, fluid, and deeply grounded in the Land, relationships to place/Land and all its inhabitants, culture, spirituality, and the cosmos (Chilisa, 2012; Datta, 2018; McGregor et al., 2018; Sutherland & Henning, 2009; Wilson, 2008). It is also socially constructed and experiential (Coburn, 2013; Kovach, 2010; Sutherland & Henning, 2009). Indigenous research has been here since time immemorial but, in recent historical times, has been enduring violent oppression at the hands of Western European colonizers, North American settlers, and social classes in positions of power (Bishop 2015; Chilisa, 2012; Simpson, 2017; Smith, 2012; Tuck & Yang, 2012; Wilson, 2008). Since the 1990s, Indigenous research has seen a more collaborative resurgence attributed to the hard work of Indigenous scholars to reclaim their own voices within academic fields (Chilisa, 2012; McGregor et al., 2018; Wilson, 2008).

In what follows, I have drawn most deeply on the seminal works of Lawrence-Lightfoot and Davis (1997) and Wilson (2008) to establish points of connection. Their works point to similar general headings that I have chosen to use for the remainder of this chapter: context, voice/stories, relationships and processes/relational accountability. Lawrence-Lightfoot and Davis (1997) describe portraiture in a more linear and, at times, more compartmentalized manner while emphasizing intentional and continuous relationships among all components of the framework. Wilson (2008) presents the ongoing circular and permanently related aspect of all components within Indigenous research, stressing the effects and affects constantly at work. Front and center in Indigenous research are the R's: relationships, respect, responsibility, relevance, reciprocity and refusal (McGregor et al., 2018).

It seems important, though, to begin with a shared foundational principle: a rejection of the negative connotations and habits associated with traditional Western research, in favor of what Lawrence-Lightfoot and Hoffman Davis (1997) call "the search for goodness":

> The portraitist's stance is one of acceptance and discernment, generosity and challenge, encouraging the actors in the expression of their strengths, competencies, and insights. She sees the actors as knowledge bearers, as rich resources, as the best authorities on their own experience. (p. 141)

This resonates with Indigenous researchers' active resistance to a long negative history of Westernized colonial research that saw (and still sees) Indigenous peoples as subjects, or data, rather than human beings (Chilisa, 2012; Coburn, 2013; Smith, 2012). Paying attention to how Indigenous research, and research with Indigenous peoples, is conducted has become a more regular occurrence, explicitly moving away from deficit-based models and towards a consciously positive stance. As such researchers make a point of clarifying the language used when doing research, noting research is done *with* or *by* Indigenous peoples rather than *on*, *for*, or *about* them (Chilisa, 2012; Coburn, 2013; Longboat, 2008; McGregor et al., 2018; Wilson, 2008).

Neither portraiture nor Indigenous research focuses only on positives. Rather, they "work towards redressing wrongs, as well as exploring how challenges are met" (Chapman, 2005; Muccio et al., 2015). Chilisa (2012) adds that "ignoring of the wisdom and hope of the researched" has a negative effect (p. 50), while Wilson (2008) suggests that a good way to move on from the negative of the past is to foster and build new relationships, and seek harmony by maintaining the positive focus of research (p. 109). This transfers the positioning of the researchers to a collaborative stance and opens reflections about forming relationships with communities (see below). McGregor et al. (2018) highlight the positive aspect of doing research and remind researchers of the conscious effort that must happen:

> We should strive to *be* good in our relations and to *do* good in our relations. If research is conducted in the same good way, with the same spirit and intent, then we will have done research that meets a high standard of accountability to all our relations. (p. 19)

Likewise, Styres and Zinga (2013) suggest bringing our "good minds" to research (p. 293), an orientation central to portraiture as well.

On Context

Context is as significant for portraiture as it is embedded within Indigenous research. Portraiture views context as a vital component of the framework and "the human experience as being framed and shaped by the setting" (Lawrence-Lightfoot & Davis, 1997, p. 41). The setting extends well beyond the physical boundaries of the place where the portraitist is working, and seeks to include historical aspects as well as all details that surround the physical place, directing attention to the multiplicity of contexts experienced simultaneously by the participant-actors (Chapman, 2007; Lawrence-Lightfoot & Davis, 1997). As far outward as context in portraiture can go, equal attention is given to inner details or particularities acquired through sensory modes. The goal, for Lawrence-Lightfoot and Davis (1997), is to "view the context as a dynamic framework — changing and evolving, shaping and being shaped by the actors" (p. 59). Hence, context affects, and is affected by, its inhabitants and is continuously in flux. Through the process of portraiture, the portraitist is careful to assemble a narrative that reflects the breadth of the context as well as the flux as experienced by the actors.

In Indigenous research context holds deep roots. Much as in portraiture, context matters and is intricately linked with knowledge, which itself "is held in the relationships and connections formed with the environment that surrounds us" (Wilson, 2008, p. 87). Context is the environment, the Land. Those relationships and connections to the Land inform practice, foster the learning of contextualized knowledge, and the birth of new relationships (Doucet, 2018; Hatcher et al., 2009; Styres, 2011; Wilson, 2008; Zinga & Styres, 2011). Therefore, context is dynamic and more so, it is an entity that is purposefully brought into every action, thought, and word. As Wilson (2008) states, "we contextualize everything that we do, and we do that contextualization in a conscious way" (p. 103). Similar to portraiture, context in Indigenous research includes a broader dimension. The difference is found within the deep relationships and connections Indigenous peoples maintain with the environment. There is a historical aspect of context (McGinty & Bang, 2016; Simpson, 2017; Smith, 2012; Tuck & Yang, 2014) as well as a spiritual, and emotional connection and energy (Datta, 2015, 2018; Sutherland & Henning, 2009; Wilson, 2008). This spiritual connection extends to the cosmos in the form of meanings and stories, ancestral and shared, relevant and connected to today (Kimmerer, 2013; Simpson, 2017; Wilson, 2008). Further, context, the Land, holds its own laws, its own way or regulating itself (Anker, 2017; Kimmerer, 2013). Context within Indigenous research is vibrant, living, and current.

Being outside, in place, on the land, is an essential element of Indigenous knowing. In her recounting of one of Nanabush's lessons, Simpson (2017) puts it clearly, "if you want to learn something, you need to take your body onto the land to do it" (p. 165). While piloting a pedagogy of Land, Zinga and Styres (2011) give themselves the goal to be on the land as much as possible (p. 64). They managed to open and maintain this connection and, although not without challenges, monitor learning *with* the land. In

Indigenous ways of knowing, the process of learning as it unfolds over time is equally, if not more, important than the knowledge initially sought (Hatcher et al., 2009; Kimmerer, 2013; Smith, 2002; Wilson, 2008). This highlights the value of portraiture as an *in* situ methodology, based on the premise "that human experience has meaning in a particular social, cultural, and historical context – a context where relationships are real, where the actors are familiar with the setting, where activity has purpose, where nothing is contrived" (Lawrence-Lightfoot & Davis, 1997, p. 43). Furthermore, portraiture, being modeled so closely on ethnography, develops a portrait over a "sustained period of time" during which relationships are formed and extensive data is gathered through watching, listening and interacting with the actors (Lawrence-Lightfoot & Davis, 1997, p.12). It is here that I believe ecoportraiture can do justice to this slow growth of understanding through close relationships between humans and the Land.

In their description of ecoportraiture, Timmerman and Piersol (2013) point out that shifting how we approach education and research, by including the more-than-human, "means making time and space to listen to these pre-existing voices" (p. 5). Instead of reproducing a colonial view of place as the backdrop to human concerns, we can come to see context as a web of living relationships that involve Land, stories, humans and other-than-humans. The challenge for ecoportraiture as it pushes beyond portraiture and responds to Indigenous ways of knowing then becomes to communicate the essence of these relationships effectively; in a sense, to speak "nature" or "Land." Nature does talk (Anker, 2017; Kimmerer, 2013). Through her work looking at Land and Law, Anker (2017) addresses the idea of nature speaking and concerns about superimposing human thought onto nature (pp. 195–196). She notes that such binary positioning ("thinking" humans as distinct from "unthinking" nature) is not well grounded (p. 196). Timmerman and Piersol (2013) remark on the same topic and note that the first thing we need to do is listen (p. 5). It may be that ecoportraiture holds the potential to address spirituality in connection to Land through its phenomenological and hermeneutic roots, but this remains a tricky area for non-Indigenous researchers to tackle.

Particular to both portraiture and Indigenous research is the notion that the researcher embraces subjectivity in full by being visible, present in the research. Objectivity and the distancing of the researcher signals an incongruous and highly colonizing stance and attitude towards Indigenous communities (Smith, 2012). It is incompatible with the relational traditions and understandings that characterize Indigenous research (Wilson, 2008). The positioning of the researcher denotes a humbled non-expert point of view that fosters close connections with the "data" and its context (Doucet, 2018; Styres & Zinga, 2013). It also highlights the relationships that are there and the ones that have grown out of the research process (Wilson, 2008). McGregor et al. (2018) provide clear imagery for this positioning, describing the researcher as "nested in concentric circles of relationships" (p. 11). Indigenous research is undertaken with the understanding that the presence of the researcher will be evident throughout the research.

Portraiture views the positioning of the researcher within the context as a purposeful, explicit, yet measured event that demands constant repositioning as the research progresses, generally in order to move towards a more subdued presence (Lawrence-Lightfoot, 2005; Lawrence-Lightfoot & Davis, 1997). The researcher modulates her presence and interactions while ensuring that she does not become the focal point through which the research will be developed. However, the presence of the researcher within the research is known to everyone involved. This shift in the stance of the researcher leads to a direct common experience with the actors as a result of, and contributing to, relationships being developed through the process of portraiture (Chapman, 2005; Gaztambide-Fernández, Cairns, Kawashima, Manna, & VanderDussen, 2011; Lawrence-Lightfoot & Davis, 1997). The idea that the researcher sits in a learner position, getting to know the new context, resonates with the Indigenous belief that context is a source of knowledge and one's mere presence already starts to establish new relationships.

On Voice/Stories

In portraiture, voice, both that of the researcher and those of the participants, is fundamental to building the narrative and creating the portrait. Voice is as pivotal in Indigenous research and ecoportraiture. Rather than list the many evident connections the notion of voice generates, the focus of this section is to reflect on the similarity of broad overarching ideas. Wilson (2008) reminds us that, from an Indigenous standpoint, "there should be no need for us to constantly justify, validate or change our work in order to fit foreign research paradigms" (p. 127). As such, voice is discussed with careful attention to the distinctiveness of portraiture and Indigenous research, along with some thoughts on how it may be present within ecoportraiture.

The core feature of portraiture, and what sets it apart from ethnography (aside from the development of an artistic piece as a product) is the intentional actions of the portraitist in listening *for* a story, rather than *to a* story. Lawrence-Lightfoot and Davis (1997) describe the act of listening *for* a story as listening for voice and "timbre, resonance, cadence, and tone [...], their message and their meaning" as well as "observ[ing] very closely, watching for the way in which the actors' movements and gestures express and communicate what they mean" (p. 99). This implies responsibility on the part of the researcher, who is once again in the position of a learner. Whereas listening *to* a story is a receiving act, listening *for* a story is an engaged interaction (Cope, Jones, & Hendricks, 2015).

Elders in Archibald (2008) also view listening as a purposeful activity. They stress that learning *how* to listen is a skill, requiring one to "not talk," to start with, and "the concomitant involvement of the auditory and visual senses, the emotions, the mind, and patience" (p. 76). From her experience with Indigenous Elders, Belczewski (2009) notes the need to listen first, then talk, then listen again (p. 198). Through an ecoportraiture lens, Blenkinsop, Affifi, Piersol, and Sitka-Sage (2017) further stress the necessity for listening first, drawing attention to the manner in which colonization suppresses other-than-humans' voices (p. 363). Listening requires more than just receiving the story, it involves processing. In French, we say "écoute avec tes deux oreilles," listen with both ears, one listening, the other one thinking. As Wilson (2008) explains, "your job is to listen, to internalize, and to be aware." (p. 134). This interactive approach to listening is common to portraiture, ecoportraiture, and Indigenous research. Within Indigenous research, listening works to allow interlocutors to build upon each other's thoughts (Wilson, 2008) and is a participatory act (McGregor et al., 2018). Listening is carefully woven throughout the methodology of portraiture, with the aim of helping the researcher arrive at a point where communicating the essence of what is experienced is possible.

Listening to *non-verbal* means of communication, whether as part of human culture or within the broader life of the Land, is as significant within portraiture and ecoportraiture as it is within Indigenous research. Much of this communication takes places through metaphors (physical metaphors, non-verbal details). Lawrence-Lightfoot and Davis (1997) note the importance of metaphors by explaining that they "act as symbols pointing to larger phenomena that will emerge as significant and be developed more fully later on in the narrative" (p. 55). Of equal importance is the need to listen for silences and notice changes or the lack thereof (p. 57). The portraitist attempts to be attuned to all aspects of context by being in place with purpose and intent; it is the work of the portraitist to do *deep listening* (Hampsten, 2015), to listen actively for "words or phrases that resonate with meaning and symbolism, sometimes representing the central core of institutional culture or the dominant dimension of a life story" (Lawrence-Lightfoot & Davis, 1997, p. 198). In discussing relationships with " other species, beings, and animate processes," Blenkinsop et al. (2017) reflect tenets of ecoportraiture by proposing that "by working at the level of metaphor, teachers might begin to re-orient their pedagogy away from the dominant root metaphors that sustain colonial relationships" (p. 363). This seems like important guidance for researchers as well.

In Indigenous research, relationality to land and to community is often expressed though stories (Archibald, 2008; Penak, 2018; Styres, 2011). The use of metaphors is omnipresent within stories; "one word is like a zip file, zip disk that crunches all this information into it" (Wilson, 2008, p. 112). As such, metaphors play a significant role for the researcher, helping with the construction of knowledge and deepening understandings. The end product, the story or the portrait, reflects the effort of the storyteller/portraitist to carefully select knowledge expressed through metaphors, to convey layers and patterns of information (Anderson, 2011; Lawrence-Lightfoot & Davis, 1997; Wilson, 2008). The final piece will inevitably include metaphors, as Wilson (2008) points out, because they "allow[s] the audience an easier way to form a relationship with something that may be abstract" (p. 124).

Michael Wilson (2005) points to Indigenous uses of symbolism beyond metaphors: "memories can be encoded in non-verbal communications such as gestures, adornment, and other symbolic media, as well as the cultural landscape" (para. 4). Portraiture has the characteristic of being malleable enough to work creatively with metaphor, as in Dixson, Chapman and Hill (2005) where Dixson mixed portraiture and jazz (p. 21). Ecoportraiture is equally malleable and, in addition, opens the researcher to understandings of beings in the natural world "as having a voice, living in fertile communities, performing culture, or enacting agencies at once unique, wonderful, and, in many ways, incomprehensible to us" (Blenkinsop et al., 2017, p. 364). This acceptance of metaphors and other signs linked to encoded memories opens the door to multiple modes of knowledge transmission and data collection for interpretation. It possesses the flexibility to bring art, science, and the world being portrayed even closer than first imagined.

On Relationships

References to relationships have permeated the previous sections as they are central to each methodology. It is clear that for these methodologies to be enacted, relationships need to happen from the moment the project emerges as an idea.

In portraiture, relationships are essential, dynamic, and demand reciprocity and responsibility from the researcher (Lawrence-Lightfoot & Davis,

1997). The framework of portraiture means that the researcher enters into a long-term relationship with the participants, at times chosen because of already existing connections (Chapman, 2005; Hampsten, 2015; Muccio et al., 2015). Previous connections may transform and evolve as a result of the research process, offering new insights for the researcher and the participants (Chapman, 2005). Relationships between components of the methodology come into play in such a manner that, even with familiar participants, or with unknown but very relatable participants, the mechanisms for ensuring quality work are created and act to counter potential biased results.

In portraiture, relationships are purposeful in nature and will vary in degrees of depth independent of time spent together (Gaztambide-Fernández et al., 2011; Lawrence-Lightfoot & Davis, 1997). Beneficence and ethical matters are addressed through deliberate attention to the reciprocity of the relationships. This results in care, dedication, and respect remaining central elements. A sign of reciprocity lies in the sharing of the portrait with the actors and seeking feedback, a conscious strength-focused process that contributes to improved data (Anderson, 2011; Lawrence-Lightfoot & Davis, 1997; Quigley et al., 2015). The relationship with the actors includes an exchange of thoughts and adjustments to the portraits that occurs frequently throughout the process. The resulting narrative is a testament to the relationships that evolved during the project. Lawrence-Lightfoot and Davis (1997) wish to ensure that the actors of the portraits feel *seen* (p. 5). A similar feeling is shared in Indigenous research, where it is a web of relationships that is sought to be *seen* by the receiver of the information, not necessarily in written form (Wilson, 2008).

Relationships, or relationality, are part of the multiple R's of Indigenous research: relationships, respect, responsibility, relevance, reciprocity and refusal (Chilisa, 2012, McGregor et al., 2018; Styres & Zinga, 2013; Wilson, 2008). As Wilson (2008) explains, "we *are* the relationships that we hold and are part of" (p. 80). This means that Indigenous research seeks ways of working that foster relationships (narratives, talking circle, participatory action research), with people and/or places, and focuses on a co-construction of knowledge (Styres, 2011; Wilson, 2008). Within Indigenous research, reciprocity is also relational in form. In this sense, naming participants within the research (only with permission of the person and in consideration of the knowledge) honors holders of knowledge (Wilson, 2008). Reciprocity, as part of relational accountability, involves a commitment to, and care of, the relationship with the participant as well as making sure the research benefits the community (Chilisa, 2012; Doucet, 2018; Wilson, 2008) – more on this below.

All of the above considerations apply to research involving Land as a living web of relationships. Two important themes from the broader portraiture literature are that relationships with the researcher and with the setting often both precede and follow the research (Chapman, 2005, p. 36; Ingman, 2017, p. 342), and that the portraitist is especially interested in particular kinds of expertise or wisdom that emerge from long relationships with a community or place (e.g. Timmerman, this volume). Thus ecoportraiture may place special emphasis on seeking out the stories and viewpoints of Elders and others who have a long continuous relationship with the Land, and also pay attention to the researcher's own roots in Land and her developing relationship with the particular places and more-than-human actors foregrounded in the research. Moreover, as emphasized by Chambers (2006), an empathetic regard can be, and should be, extended to Land as a means to gain a deeper understanding of its subjective reality (see also Lawrence-Lightfoot & Davis, 1997, p. 147). From an Indigenous perspective of relationality, Simpson (2017) calls such empathetic regard *mirroring*, stating that "part of being in a meaningful relationship with another human being is recognizing who they are, it is reflecting back to them their essence and worth as a being" (p. 180). A powerful example of this beyond the solely human and in the context of a specific place is Kimmerer's (2013) account of Lake Onondaga's response to a nearby industrial plant. The extensive and detailed description rendered by Kimmerer (2013) is akin to the portrait of a community, an actual "ecoportrait," of the beings whose lifeways have been devastated by the plant. Ecoportraiture as a methodology brings an empathetic regard, or mirroring relationship, to bear for the purpose of highlighting voices such as the ones in Lake Onondaga.

On Processes/Relational Accountability

The centrality of relationship affects the standards to which portraiture and Indigenous research hold themselves. Key concepts here are integrity/authenticity and transformation.

Both portraiture and Indigenous research do away with notions of validity and reliability as they are traditionally viewed within Western research, with its strong bias towards generalizability and reproducibility. In portraiture, the focus shifts to authenticity, which is present in the rigorous attention to details and as a result of the relationships developed throughout the process (Gaztambide-Fernández et al., 2011; Lawrence-Lightfoot & Davis, 1997). Constant member-checking strategies all along the research project contributes to developing an authentic portrait (Lawrence-Lightfoot & Davis, 1997;

Quigley et al., 2015). This extends to *tests* designed to measure for accuracy and to ensure that the researcher is not falling into the trap of seeing what she wants to see. Those tests – resonance, coherence and necessity – act as a set of accountability parameters, where member-checking is paired with analysis and re-analysis of the information collected and assembled into the portrait through a process called *assemblage* (Lawrence-Lightfoot & Davis, 1997). Through assemblage, every piece of the narrative is checked and validated against all others (p. 245). Using Aristotle's whole to part analogy, the authors note "the structural union of the parts is such that, if any one of them is displaced or removed, the whole will be disjointed and disturbed" (Lawrence-Lightfoot & Davis, 1997, p. 268). Validity also comes in the form of a credible story, reflecting the presence and work of the researcher in-situ (Gaztambide-Fernández et al., 2011; Lawrence-Lightfoot & Davis, 1997). The process is rigorous and meticulous, and extends past the actors within the project, to the audience who also needs to experience this resonance. Lawrence-Lightfoot & Davis (1997) conclude "when all these resonances — the researcher's, the actors', the audience's — echo through the piece, we speak of the portrait as achieving authenticity" (p. 260). The portrait conveys the essence.

Indigenous research looks at authenticity, integrity and credibility as a means to ensure what Wilson refers to as relational accountability (Chilisa, 2012; Wilson, 2008). Nestled in a relational mindset, credibility is achieved by the positioning of the researcher within the research, and by the researcher exercising reflexivity through rigorous reflection on all aspects of their positionality prior to and during the research (Belczewski, 2009; Chilisa, 2012; Doucet, 2018; Styres, 2011). Furthermore, the researcher must have and display integrity, and demonstrate trust and responsibility towards the research itself (Wilson, 2008). Because the research is based on fostering participation, the analysis of the information is done within the research process with the participants through a process called *encircling* (Wilson, 2008). The voice of the participants must be reflected in a true manner with the analysis, and the research must reflect the relationships at work (hence naming participants by name) (Chilisa, 2012; Wilson, 2008). The multiple R's of Indigenous research provide guidelines for disrupting the influences of colonialism (Chilisa, 2012; McGregor et al., 2018). As Wilson (2008) explains, "even though we don't need externally imposed measures or tests of whether or not something is 'true', we have our own ways of ensuring this" (p. 102). In order to understand Indigenous research, one must understand the importance and the depth of the relationships, and relationship networks, always in movement. This is all encompassing. Wilson (2008) states, "the whole of the paradigm is greater than the sum of its parts" (p. 70). He also emphasizes that the essence of relationality lies in the realization that the context and the content are never separate from each other (p. 99).

This last observation holds true for portraiture and ecoportraiture as well. We might say that ecoportraiture expands both content and context to include the more-than-human world, in the same way Indigenous research might expand them to include Land, ancestors, and the world of spirit. In all of these cases, decisions about what is most relevant and important to include are guided by the voices of the participants themselves.

The goal, in all of these forms of research, is not simply to describe a setting or community. Working from a perspective of goodness holds transformative potential for participants/actors, researchers, and audiences alike. In the construction of new knowledge, new insights can be formulated that may lead to transformation. This is an outcome that both portraiture and Indigenous research identify as part of the process. Wilson (2008) suggests that Indigenous research has an "axiology that research is not seen as worthy or ethical if it does not help improve the reality of research participants" (p. 37); others position it as a means. It is clear, however, that although transformation occurs, the aim is not to impose it as an external ideal, formulated outside the research setting. A way to put it might be to use the word "recognize." Transformations may happen because a process of recognition *within the research setting* is set in motion.

Lawrence-Lightfoot and Davis (1997) point to transformation as a feature of portraiture methodology because of the intentional probing in all directions throughout the research. They suggest that portraiture calls attention to deliberate acts of intervention and social transformation, establishing it as "provocative work that can disturb the natural rhythms of social reality and encounter" and that carries "huge ethical responsibilities" (p. 11). Portraiture thus has the ability to be a tool of collective agency, to transform relationships and "erode the status quo" (Chapman, 2007, p. 159). With an aim to transform participants, researchers and readers (Cope et al., 2015; Rogers & McLendon, 2015) alike, portraiture offers another means to reframe perspectives and understandings, one relationship at a time. This process may enable all of these audiences to see beyond their immediate reality and may influence how they view themselves in the future (Lawrence-Lightfoot & Davis, 1997, p. 35).

Indigenous research strives to benefit the community or give back to the community (Anker, 2017; Archibald, 2008; Wilson, 2008). Transformation, here, is not just a feature of the methodology, it is a relational goal from the onset. Because knowledge is relational, and thus continuously being constructed, "the aim of the research is to come to a consensus among researcher and subjects on a construction that is better informed than it was before" (Wilson, 2008, p. 37). The transformation that may result from this shared endeavor affects all participants involved in the research, including the researcher (Archibald, 2008; Wilson, 2008). Though the construction of knowledge extends to the community, Indigenous research does not seek to generalize this knowledge. Instead, the researcher, as storyteller, puts all elements of the research in place, based on an awareness of their listeners (audience), and lets the listeners make their own connections (Hatcher et al., 2009; Wilson, 2008). It is understood that what works in one community cannot be assumed to apply in another, as each Indigenous community holds its own knowledge (Smith, 2012; Wilson, 2008). Knowledge is shared but never imposed. The audience, each in their own time, will make the connections they need.

These ethical commitments and concerns are found in ecoportraiture as well. By foregrounding the relationships between humans and the more-than-human world, not in the abstract but in the concrete particulars of the research setting, ecoportraiture has the potential to transform how the natural world is understood, incorporated, recognized, and listened to – not only in the research process itself, but in the ongoing understandings and practices of the participants. Like Indigenous research, it may lead to "a construction that is better informed than it was before" – a set of relationships that have gone some way to becoming "decolonized" (Blenkinsop at al., 2017).

Decolonizing Portraiture

While I worked on this chapter, a multitude of sticky notes and strings of scotch-taped papers graced my walls, coffee tables, living room, and dining room. One of them, a purple sticky note ripped in half, stared at me for a long time. It had the words *Can portraiture help address colonialism?* scribbled on it. It was a question that sat there for the better part of the writing process. I understand the difference between the Land education currently underway in my school district and the Land education we need to aim for. There is a significant gap. A gap portraiture could show. A gap ecoportraiture, perhaps, can help to bridge.

I have no definitive answer to the sticky note question. Or rather, as I discovered on my daily walks, I have two answers. Neither one of them is convincing; one answer is phrased *yes, but how…*, and the other is phrased *no, but why not….* I am cognizant of the transformative capacity of portraiture, and I've kept asking myself, "For the research project I undertake with

Land education, what transformation should happen?" Or, should I be satisfied with considering what transformation *might* happen? What about settler colonialism? Should that be purposely addressed by questions, or observations, or in any other way during my research project?

Wilson (2008) warns against trying to decolonize methodologies because that entails attempting to fit one paradigm into another (p. 39). But, he does see aspects of some research methods as being compatible with decolonization. Smith (2012) and Tuck and Yang (2012) caution that decolonization is a deep and lengthy process, not just a tag word. Pairing portraiture directly with colonialism, settler colonialism and decolonization, is an intricate task as these terms involve complex foundational knowledge. There is a risk that knowledge centered around these terms becomes packaged into a catch phrase in the same manner decolonization has entered the educational lexicon (Tuck & Yang, 2012, p. 2). There are, however, some subtle adjustments to portraiture that may help make it more effective as a decolonizing methodology, and may contribute to the further development of ecoportraiture.

Addressing the structuring of dominant cultures' speech is one of the ways towards decolonization. Written at a different time, the expression "giving voice" is used often throughout *The Art and Science of Portraiture* by Lawrence-Lightfoot and Davis (1997). It is also echoed in research where researchers speak of students and the notion of giving them voice (Anderson, 2011; Chapman, 2005). Some of the communities referred to are viewed as marginalized and thus, in the researchers' view, portraiture contributes to empowering them by giving them a voice. Styres and Zinga (2013) question this concept: "the idea of empowering a community is fraught with contradictions as the idea that one group or individual can empower another group or individual is [. . .] based on implicitly assumed inequalities grounded in colonial relations" (p. 297). Nasuʔkin Pierre expressed a very similar thought a few years ago, "Sylvie, you don't give voice to Indigenous students," indicating that this was not appropriate, nor accurate, language to use (personal communication, June 2015). Indeed, portraiture as a co-creation project is communal work and thus, should be viewed, thought, and expressed as such.

Similarly problematic is the use of the terms "discovery/discovering," which betray a lack of understanding that knowledge has always been there. Although well-meaning, the language of thirty years ago may no longer be suitable. Semantics matter greatly and contribute to reproducing settler colonial language in a place and Land education context (Tuck et al., 2014). The alternative is to view the acquisition of knowledge through a relationality lens; as Wilson (2008) explains, "knowledge cannot be owned or discovered but is merely a set of relationships that may be given visible form" (p. 127).

In this light, knowledge is available if we are willing to establish right relationships. Instead of "discovering," maybe researcher and participants are "knowledge building."

Lastly, portraiture may help expand on the description of colonialism, raise awareness, and then perhaps transform, by bringing concrete current examples into the research process for examination and discussion. In addition to the ecoportrait, a literature review could be established, for the purpose of providing co-researchers, participants and readers with relevant background knowledge – for example, to consider Land education as it might look in the hands of Indigenous educators, and to bring to light some of the tensions that exist in relation to current practices. It could also define some of the terms and relationships the reader may need to be aware of, in a similar style to how Tuck and Yang (2012) describe decolonization (p.2), and in the same manner that Wilson (2008) provides a section of his book to describe terms for the reader (pp. 32–42). Because this is complex, Elders could be invited into the process to enact member-checking, and to infuse guiding thoughts.

Throughout their book, Lawrence-Lightfoot and Davis (1997) add artwork in progressive states. The display of the progressive states of artwork reveals the underlying depth of the final project. This is how I view my suggested adaptation of portraiture. When looking at portraiture and settler colonialism in the context of Land, it may be helpful to overlay frameworks of understandings for the researcher, the co-researcher and the reader as part of the process of working towards the final portrait, which itself will reveal the richness of layers emanating from the research project.

Bringing the voices of the more-than-human world into the process, as ecoportraiture entails, connects with all of these suggestions. To develop knowledge in collaboration with the natural world, and to deliberately seek out Indigenous understandings of Land and relationships with the more-than-human (Simpson, 2017; Sutherland & Henning, 2009), would add important layers to the portrait. The inclusion of the more-than-human would go beyond what English alone is capable of expressing, and honor Indigenous community members' (and non-Indigenous members approved by the community) invaluable knowledge as developed through ethnobothany, archeology, trapping/hunting and other skilled, long-term relationships with the natural world.

Blenkinsop et al. (2017) demonstrate how layers of colonialism have silenced the natural world, and how colonial conceptualizations of the natural world maintain the status quo. Ecoportraiture may therefore offer an opportunity to resist colonialism explicitly, not only by listening to new voices such as "the specific skills, talents, and ways of being of *this crow*, *those earthworms*, or *that bumblebee*" (p. 357), but exploring how Indigenous ways of knowing offer greater space for those voices to be heard.

In their work pairing Western science and Indigenous knowledge, McGinty and Bang (2016) conclude with a description of students interacting with the audience to share their results (p. 475). The interactive ending is appealing as a means to share with audiences because it brings the co-creation process into a co-sharing presentation, including live exchanges with the audience. Students and researchers remain in the real world, sharing, inspiring, and hopefully transforming. Similarly, when working with ecoportraiture, it might be interesting to go a step beyond and find an interactive way to share the written portrait. This would also honor Indigenous knowledge sharing, which is based on oral traditions rather than written texts (Balanoff & Chambers, 2005; McGregor et al., 2018). The choice of art medium would likely emerge from the nature of the place and Land education project. Finley (2018) offers up to nine suggestions for art-based research genres (dance, music, variation of visual arts, drama) and variations of narratives (short stories, novels, films) (p. 572). Teachers and students can come up with a million additional ways.

Adapting and extending ecoportraiture to embrace Indigenous ways of knowing is a dynamic process. Portraiture methodologies are particularly suited for such pedagogies/ways of knowing because of their strong emphasis on context, relationships and expression (voice). These are foundational elements: *they are us* – to borrow Wilson's (2008) explication. At a time when my own country of Canada is trying to understand what reconciliation might look like, and when Indigenous peoples themselves are calling for an end to "eco-ignoring" and "eco-avoidance," ecoportraiture offers a methodology that is not only provocative, it responds to some of our most pressing contemporary issues. It's an appealing choice.

References

Anderson, L. M. (2011). Toward professional integration in the humanities: One teacher-researcher's experience with portraiture. *Arts and Humanities in Higher Education: An International Journal of Theory, Research and Practice*, *10*(1), 103–119. doi: https://doi.org/10.1177/1474022210389143

Anker, K. (2017). Law as…forest: Eco-logic, stories and spirits in indigenous jurisprudence. *Law Text Culture*, *21*, 191–213. Retrieved from https://heinonline.org/HOL/LandingPage?handle=hein.journals/lwtexcu21&div=13&id=&page=

Archibald, J. A. (2008). *Indigenous storywork: Educating the heart, mind, body, and spirit.* Vancouver: UBC Press.

Balanoff, H., & Chambers, C. (2005). Do my literacies count as literacy. *Literacies*, *6*(19), 18–20. Retrieved from http://literacies.ca/literacies/6-2005/pdf/balanoff+chambers.pdf

Bang, M., Curley, L., Kessel, A., Marin, A., Suzukovich III, E. S., & Strack, G. (2014). Muskrat theories, tobacco in the streets, and living Chicago as Indigenous land. *Environmental Education Research*, *20*(1), 37–55. doi: https://doi.org/10.1080/13504622.2013.865113

Belczewski, A. (2009). Decolonizing science education and the science teacher: A White teacher's perspective. *Canadian Journal of Science, Mathematics and Technology Education*, *9*(3), 191–202. doi: https://doi.org/10.1080/14926150903118326

Bishop, A. (2015). *Becoming an ally: Breaking the cycle of oppression in people.* London: Zed Books.

Blenkinsop, S., Affifi, R., Piersol, L., & Sitka-Sage, M. D. D. (2017). Shut-up and listen: Implications and possibilities of Albert Memmi's characteristics of colonization upon the "natural world". *Studies in Philosophy and Education*, *36*(3), 349–365.

Bowers, C. A. (2008). Why a critical pedagogy of place is an oxymoron. *Environmental Education Research*, *14*(3), 325–335. doi: https://doi.org/10.1080/13504620802156470

Chambers, C. (2006). "The land is the best teacher I have ever had": Places as pedagogy for precarious times. *Journal of Curriculum Theorizing*, *22*(3), 27–37. Retrieved from https://search.proquest.com/openview/691f422980f7ec08ece8c6aea3863ad6/1?pq-origsite=gscholar&cbl=25997

Chapman, T. K. (2005). Expressions of "voice" in portraiture. *Qualitative Inquiry*, *11*(1), 27–51. doi: https://doi.org/10.1177/1077800404270840

Chapman, T. K. (2007). Interrogating classroom relationships and events: Using portraiture and critical race theory in education research. *Educational Researcher*, *36*(3), 156–162. doi: https://doi.org/10.3102/0013189X07301437

Chilisa, B. (2012). *Indigenous research methodologies.* Los Angeles: Sage Publications.

Chrona, J. L. (2016). *First peoples principles of learning.* Retrieved from https://firstpeoplesprinciplesoflearning.wordpress.com/background-and-current-context/

Coburn, E. (2013). Indigenous research as resistance. *Socialist Studies/Études Socialistes*, *9*(1), 52–63. doi: https://doi.org/10.18740/S40017

Cope, V., Jones, B., & Hendricks, J. (2015). Portraiture: A methodology through which success and positivity can be explored and reflected. *Nurse Researcher*, *22*(3), 6–12. doi: https://doi.org/10.7748/nr.22.3.6.e1311

Datta, R. K. (2015). A relational theoretical framework and meanings of land, nature, and sustainability for research with indigenous communities. *Local Environment*, *20*(1), 102–113. doi: https://doi.org/10.1080/13549839.2013.818957

Datta, R. K. (2018). Rethinking environmental science education from indigenous knowledge perspectives: An experience with a Dene First Nation community. *Environmental Education Research*, *24*(1), 50–66. doi: https://doi.org/10.1080/13504622.2016.1219980

de Grandpré, S. (2017). Embedding the British Columbia First Peoples Principles of Learning within the curriculum. Unpublished paper, University of Lethbridge.

Derby, M. W. (2015). *Place, being, resonance: A critical ecohermeneutic approach to education.* New York, NY: Peter Lang.

Dixson, A. D., Chapman, T. K., & Hill, D. A. (2005). Research as an aesthetic process: Extending the portraiture methodology. *Qualitative Inquiry*, *11*(1), 16–26. doi: https://doi.org/10.1177/1077800404270836

Doucet, A. (2018). Shorelines, seashells, and seeds: Feminist epistemologies, ecological thinking, and relational ontologies. In F. Dépelteau (Ed.), *The Palgrave handbook of relational sociology* (pp. 375–391). London: Palgrave Macmillan.

Finley, S. (2018). Critical arts-based inquiry: Performance of resistance politics. In N. Denzin & Y. Lincoln (Eds.), *The SAGE handbook of qualitative research* (pp. 561–575). Thousand Oaks, CA: Sage.

First Nations Education Steering Committee. (2008). *First peoples principles of learning.* Retrieved from http://www.fnesc.ca/wp/wp-content/uploads/2015/09/PUB-LFP-POSTER-Principles-of-Learning-First-Peoples-poster-11x17.pdf

Gaztambide-Fernández, R., Cairns, K., Kawashima, Y., Menna, L., & VanderDussen, E. (2011). Portraiture as pedagogy: Learning research through the exploration of context and methodology. *International Journal of Education & the Arts*, *12*(4), 1–29. Retrieved from http://www.ijea.org/v12n4/v12n4.pdf

Gehl, L. (2011). *Ally bill of responsibilities.* Retrieved from http://www.lynngehl.com/uploads/5/0/0/4/5004954/ally_bill_of_responsibilities_poster.pdf

Greenwood, D. A. (2009). Place, survivance, and white remembrance: A decolonizing challenge to rural education in mobile modernity. *Journal of Research in Rural Education (Online)*, *24*(10), 1–6. Retrieved from http://citeseerx.ist.psu.edu/viewdoc/download?doi=10.1.1.609.1156&rep=rep1&type=pdf

Gruenewald, D. A. (2003). Foundations of place: A multidisciplinary framework for place-conscious education. *American Educational Research Journal*, *40*(3), 619–654. doi: https://doi.org/10.3102/00028312040003619

Gruenewald, D. A. (2004). A Foucauldian analysis of environmental education: Toward the socioecological challenge of the Earth Charter. *Curriculum Inquiry*, *34*(1), 71–107. doi: https://doi.org/10.1111/j.1467-873X.2004.00281.x

Hampsten, K. (2015). Challenges in working with portraiture. *Journal of Applied Communication Research*, *43*(4), 468–471. doi: https://doi.org/10.1080/00909882.2015.1083605

Hampton, R., & DeMartini, A. (2017). We cannot call back colonial stories: Storytelling and critical land literacy. *Canadian Journal of Education*, *40*(3), 245–271. Retrieved from https://cje-rce.ca/wp-content/uploads/sites/2/2017/10/9.-2543-Hampton-Sept-18.pdf

Harrison, N., & Greenfield, M. (2011). Relationship to place: Positioning Aboriginal knowledge and perspectives in classroom pedagogies. *Critical Studies in Education*, *52*(1), 65–76. doi: https://doi.org/10.1080/17508487.2011.536513

Hatcher, A., Bartlett, C., Marshall, A., & Marshall, M. (2009). Two-eyed seeing in the classroom environment: Concepts, approaches, and challenges, *Canadian Journal of Science, Mathematics and Technology Education*, *9*(3), 141–153. doi: https://doi.org/10.1080/14926150903118342

Hursh, D., Henderson, J., & Greenwood, D. (2015). Environmental education in a neoliberal climate. *Environmental Education Research*, *21*(3), 299–318. doi: https://doi.org/10.1080/13504622.2015.1018141

Ingman, B. C. (2017). Bigger experiences and the meaning of adventure. *Journal of Adventure Education and Outdoor Learning*, *17*(4), 338–352. doi: https://doi.org/10.1080/14729679.2017.1291356

Johnson, J. T. (2012). Place-based learning and knowing: Critical pedagogies grounded in Indigeneity. *GeoJournal*, *77*(6), 829–836. doi: https://doi.org/10.1007/s10708-010-9379-1

Kimmerer, R. W. (2013). *Braiding sweetgrass: Indigenous wisdom, scientific knowledge and the teachings of plants.* Minneapolis, MN: Milkweed Editions.

Kovach, M. (2010). Conversational method in indigenous research. *First Peoples Child & Family Review*, *5*(1), 40–48. Retrieved from https://fncaringsociety.com/sites/default/files/online-journal/vol5num1/Kovach_pp40.pdf

Kovach, M. (2018). Doing indigenous methodologies: A letter to a research class. In N. Denzin & Y. Lincoln (Eds.), *The SAGE handbook of qualitative research* (pp. 214–234). Thousand Oaks, CA: Sage.

Lawrence-Lightfoot, S. (2005). Reflections on portraiture: A dialogue between art and science. *Qualitative Inquiry*, *11*(1), 3–15. doi: https://doi.org/10.1177/1077800404270955

Lawrence-Lightfoot, S., & Davis, J. H. (1997). *The art and science of portraiture.* San Francisco, CA: Jossey-Bass.

Longboat, C. (2008). Ethical space in the intellectual terrain: A cultural perspective. *Canadian Journal of Native Education*, *31*(1), 72–83. Retrieved from https://search.proquest.com/openview/44e9d9151b0a2597861c77d0153170bc/1?pq-origsite=gscholar&cbl=30037

Louv, R. (2012). *The nature principle: Reconnecting with life in a virtual age.* Chapel Hill, NC: Algonquin Books.

McGinty, M., & Bang, M. (2016). Narratives of dynamic lands: Science education, indigenous knowledge and possible futures. *Cultural Studies of Science Education*, *11*(2), 471–475. doi: https://doi.org/10.1007/s11422-015-9685-5

McGregor, D., Restoule, J. P., & Johnston, R. (Eds.). (2018). *Indigenous research: Theories, practices, and relationships.* Toronto, ON: Canadian Scholars' Press.

McInerney, P., Smyth, J., & Down, B. (2011). 'Coming to a place near you?': The politics and possibilities of a critical pedagogy of place-based education.

Asia-Pacific Journal of Teacher Education, 39(1), 3–16. doi: https://doi.org/10.1080/1359866X.2010.540894

Muccio, L. S., Reybold, L. E., & Kidd, J. (2015). Researcher-portraitists: An exploration of aesthetics and research quality. *International Journal of Education & the Arts, 16*(1), 1–24. Retrieved from http://www.ijea.org/v16n1/v16n1.pdf

Penak, N. (2018). A story pathway: Restoring wholeness in the research process. In D. McGregor, J. P. Restoule, & R. Johnston (Eds.), *Indigenous research: Theories, practices, and relationships*. Toronto, ON: Canadian Scholars' Press.

Podruchny, C. (2013, November 18). Indigenous history in the classroom: Four principles, four questions [Web log post]. Retrieved from http://activehistory.ca/2013/11/indigenous-history-in-the-classroom-four-principles-four-questions/

Quigley, C., Trauth-Nare, A., & Beeman-Cadwallader, N. (2015). The viability of portraiture for science education research: Learning from portraits of two science classrooms. *International Journal of Qualitative Studies in Education, 28*(1), 21–49. doi: https://doi.org/10.1080/09518398.2013.847507

Rippin, A. (2012). Eliza, Anita and me: An art investigation into using portraiture as a research method in organisation studies. *Culture & Organization, 18*(4), 305–322. doi: https://doi.org/10.1080/14759551.2012.705529

Rogers, C., & McLendon, A. (2015). Shayla: A portrait of the schooling experiences of an eastern band of Cherokee Indian youth. *Multicultural Perspectives, 17*(4), 190–197. doi: https://doi.org/10.1080/15210960.2015.1079490

Simpson, L. B. (2017). *As we have always done: Indigenous freedom through radical resistance*. Minneapolis, MN: University of Minnesota Press.

Smith, G. A. (2002). Place-based education: Learning to be where we are. *Phi Delta Kappan, 83*(8), 584–594. doi: https://doi.org/10.1177/003172170208300806

Smith, L. T. (2012). *Decolonizing methodologies*. Dunedin, New Zealand: Otago University Press.

Smitka, J. A. (2015). Exploring portraiture in teaching practice. *Art Education, 68*(5), 20–27. doi: https://doi.org/10.1080/00043125.2015.11519335

Sobel, D. (1996). *Beyond ecophobia*. Great Barrington, MA: Orion Society.

Sobel, D. (2005). *Place-based education: Connecting classrooms & communities*. Great Barrington, MA: Orion Society.

Styres, S. D. (2011). Land as first teacher: A philosophical journeying. *Reflective Practice, 12*(6), 717–731. doi: https://doi.org/10.1080/14623943.2011.601083

Styres, S., & Zinga, D. (2013). The community-first land-centred theoretical framework: Bringing a 'good mind' to Indigenous education research? *Canadian Journal of Education/Revue canadienne de l'éducation, 36*(2), 284–313. Retrieved from http://www.jstor.org/stable/canajeducrevucan.36.2.284

Sutherland, D., & Henning, D. (2009). Ininiwi-kiskānītamowin: A framework for long-term science education. *Canadian Journal of Science, Mathematics and Technology Education, 9*(3), 173–190. doi: https://doi.org/10.1080/14926150903118359

Timmerman, N., & Piersol, L. (2013, June). Storied sketches of people and place. In Presentation at the annual conference of the Canadian Society for the Studies of Education, Victoria, BC.

Tuck, E., McKenzie, M., & McCoy, K. (2014). Land education: Indigenous, post-colonial, and decolonizing perspectives on place and environmental education research. *Environmental Education Research*, *20*(1), 1–23. doi: https://doi.org/10.1080/13504622.2013.877708

Tuck, E., & Yang, K. W. (2012). Decolonization is not a metaphor. *Decolonization: Indigeneity, Education & Society*, *1*(1), 1–40. Retrieved from https://jps.library.utoronto.ca/index.php/des/article/view/18630/15554

Wilson, M. C. (2005). The "placing" of identity in nomadic societies: Aboriginal landscapes of the northwestern plains of North America. *Material Culture Review/Revue de la culture matérielle*, *62*, 7–19. Retrieved from https://journals.lib.unb.ca/index.php/MCR/article/view/18057

Wilson, S. (2008). *Research is ceremony: Indigenous research methods*. Black Point, NS: Fernwood Publishing.

Zinga, D., & Styres, S. (2011). Pedagogy of the Land: Tensions, challenges, and contradictions. *First Nations Perspectives*, *4*(1), 59–83. Retrieved from http://mfnerc.org/wp-content/uploads/2012/11/5_ZingaStyres.pdf

6. *Traversing a Mountain of Data: Ecoportraiture as Critical, Beautiful Risk*

MICHAEL DÉ DANANN DATURA[1]

> If we play for a moment with the etymology of "data" we find that it originally means "that which is given" or "that which is granted..." What is given or granted is precious and delicate, and it must be gathered with all the love and care with which we gather the fruits of the earth, careful not to do violence, careful not to expect too much, prepared to wait, prepared—dare we admit it?—for the possibility that nothing will come forth... This gathering cannot take its cue from our willingness to act without a moments notice. It must take its cue from that which is given as to how the gathering must go. There is nothing we can do to guarantee that this gift will be given. (Jardine, 1996, p. 30)

> The phenomenology of experience is particularly relevant in situations where there is evidence that certain experiences are shared but also ignored by ideological emphases in the culture. Attending to their phenomenology is one way to begin to take them seriously; and begin to reflect on the politics of their marginalization. (Zwicky, 2014, p. L109)

I. *Field Notes & Ontology: Some Thoughts on Ecologizing Portraiture*

It took all morning to walk from Pokhara to the base of the Panchase pilgrimage, the purported five-peaked seat of yogic consciousness (*panch*: five,

[1] This chapter is partially inspired by a collaborative writing experiment composed by four of the field researchers from the Maple Ridge Environmental School Project: myself, Laura Piersol, John Telford and Michael Caulkins. I would like to acknowledge their contributions, camaraderie and the invaluable feedback they provided. Stay wild, comrades.

asan: seat). The rest of the day we remained intentionally silent—traversing some fifteen hundred feet up ancient stone steps through tiered paddy fields and subtropical forest, stopping only to smoke hashish from a stone chillum and drink river water—until we reached a small village with panoramic views of the Annapurna massif. My spiritual comrade Asia, a Sri Lankan bohemian-type with an impossibly smooth London accent, turned to me then, at that most precious of moments, to share a chengyu—an idiomatic set phrase—"these two Chinese girls" had taught him on a previous trek. Something to the effect of: "One day in the mountains is a thousand years." I gazed down the range at double-peaked Machapuchare (which translates from Nepali to "Mt. Fishtail" for obvious reasons) nestled, from this angle, in the geologic caress of Mother Annapurna as a spellbinding dusk fell on its erratic slopes and plateaus. So true. Like a fool, I snap a few photos, confident I have captured the genius loci in high definition. Months later, I am holding the matte impressions in my hands, reaching for the presence of what Asia called the "divine topography" in the pixels, and finding not a trace. There are some good portraits of Asia standing in front of various pagodas and ponds, which make me smile so I keep them, but I discard all the "nature shots" from Panchase. How did I ever hope to capture the essence of such a place?

✖

As we trek up a hillside in Malcolm Knapp Research Forest, nestled in the foothills of the Golden Ears Mountains,[2] with some eighty-odd elementary students, ostensibly trying to learn math, the *chengyu* returns to me. This time with a facetious twist: "One day at the Maple Ridge Environmental School is a thousand years." So true. As one of the field researchers, I have been charged with gathering "ethnographic data" aimed at providing an in-depth understanding of pedagogical norms; that is, the social routines, shared interpretations, forms of discourse, and so on that emerge in such an immersive place-based project. Data collection methods thus far have included many of the hallmarks of qualitative research: participant observation, discourse analysis of the curriculum, semi-structured interviews and focus groups, but it strikes me that none of these are *necessarily* place-based methods. They do not, as hermeneutic scholar David Jardine suggests, "take their cue from that which is given as to how the gathering must go." Nor do they capture anything of the essence of *this* particular place beyond acting as

[2] Malcolm Knapp Research Forest and the southern part of Golden Ears Provincial Park are situated primarily within the traditional territories of the Katzie First Nation and the Kwantlen First Nation who have dwelled in *S'ólh Téméxw*, according to their *sx̱wōx̱wiyám* (oral histories), from time immemorial.

a scenic background for human subjects. One of the guiding principles of the school is to "cultivate learning in natural settings, where we listen for what the more-than-human world has to teach us." It follows that the research team ought to also cultivate methodologies attuned with and informed by the myriad beings that co-constitute this place. But how? Semi-structured interviews with a local rout of banana slugs? A thick description of the seasonal behavior patterns of various salal patches? A sampling of "participant-generated" artefacts? I often stop to sample the salmon berries—does that count as place-based research?

✖

In her description of portraiture, Sara Lawrence-Lightfoot explores what form of inquiry we might employ as educational researchers to "capture the complexity and aesthetic of human experience?" (2005, p. 5). A fine question, one I should like to ponder in greater depth, if not for the fact that in addition to gathering ethnographic data, I find myself almost constantly zipping up rain jackets, responding to a barrage of questions about fungi I cannot identify, and keeping my head permanently on safety swivel mode. We are three years into the research project and have had at least one field researcher at the school almost every day thus far. One of the original intentions of the project was to consciously disrupt the boundaries between teachers, researchers, community members, and students, such that roles would be nested within each other, within this place. I am unsure to what degree we have realized this intention—cultural norms and hierarchies are remarkably resilient—but due to our prolonged, immersive, boundary-troubling approach we have managed to accumulate a *massive* quantity of data that the research team has taken to calling, somewhat ominously, "the mountain." Trouble is, a mountain of data does not necessarily make for good research—a story line needs to emerge from the material, ideally one that takes its cue from that which is given. Traversing this mountain with any kind of accuracy or integrity seems risky on all sides: the formal and epistemological demands of conventional academe, the crisis of representation (of both human and other-than-human participants), the politics of marginalized ways of knowing. "So your research is... about teaching kids to talk with moss or something? Are you serious?" Portraiture is a methodology accustomed to blurring boundaries and challenging cultural norms, but going beyond the human will require modes of inquiry and composition informed by a different kind of ontological attunement. Let's take a walk.

✖

Every day seems like a thousand-years-worth of moments—all flowing by like river water. A young boy stomps on a ground beetle right in front of me and before I can say anything, a Kindergarten girl grabs him by the hand and says: "Don't kill bugs, they're gifts to us from God." Behind me, three older boys jump imaginary goombas in a *Nintendo*-grified world they have projected onto the landscape—apparently bored by our trek through mixed terrain that includes stands of 400-year-old trees. Up at the front of the line, the principal points out various geologic features in an attempt to foreground the "story of this place" without contradicting Katzie *sx̱wō̱x̱wiyám* (oral histories). What form of inquiry might we employ to capture the complexity and aesthetic of so many varied human experiences? And what form of inquiry might capture something of the complexity and aesthetic of the other beings that co-constitute this place? As we traverse the mountain everyone seems convinced that we are learning *something*, but it is difficult to articulate what exactly. The forest is just so full of things.

✖

My fellow field researcher and spiritual comrade, Laura Piersol, has been experimenting with ecologizing portraiture as a place-attuned methodology. Her work examines moral development in relation to the natural world by composing portraits of the evolving ethical orientations of three of the students and one of the teachers at the Maple Ridge school. Her project extends conventional conceptions of portraiture to include arts-based explorations of place informed by three eco-ethical premises:

i) entities within the natural world have intrinsic value and agency
ii) we are interdependent with each other
iii) we should therefore act with humility and caution

As her research illustrates, a key aspect to developing what she calls *eco-care* has to do with cultivating a lived practice of attending to "wild others." This is, of course, easier said than done and time intensive—amongst other, more deeply philosophical challenges. But Laura suggests if environmental researchers and educators can cultivate such practices and orientations (read: ontological attunement), they may be able to open themselves to co-creating educational experiences *with* place as a kind of co-teacher (Blenkinsop & Beeman, 2010; Blenkinsop & Piersol, 2013). What is, perhaps, even more interesting about her research is that it suggests, at times, children manage to nurture eco-care in spite of our pedagogic interventions and maintain eco-sensitivity from direct experience within sensorially-rich, biodiverse places. While this work is obviously significant in terms of its pedagogic implications for the field of environmental education, its real brilliance lies in the way in which Laura embodies a practice of *being with* both participants and "the natural world" that foregrounds inquiries into the role that other-than-human beings play in our lives. She admits the limitations of focusing primarily on "only four human voices," and yet, as a result of composing in-depth portraits over an extended period of time and thoughtfully piecing them together, she manages to traverse the mountain and capture something of the aesthetic whole of the project without relegating place to a scenic background.

✖

From *Eco care: Nurturing possibility & resistance within education* (Piersol, 2015):

> Down at the river a student turns over a rock and then calls out to me, "Laura, come quick!" I find my way through the water and crouch to see what she has found. "What is it?" she asks. There is a tiny tube made of rocks that are not much larger than a grain of sand. "Look at all the colours!" the student exclaims. Each small rock is a different shade, some pink, emerald, light purple, a miniature mosaic. It's a caddisfly house, built by the larva that lives inside with its own silk and the substrate of the stream. I like the tenacity and ingenuity of this underwater architect that has carefully crafted a strong structure of what lies around it.
>
> I decided to model my portraits after the caddisfly, collecting and gathering bits of stories here and there, piecing them together to form a narrative casing to house some tender living lessons. The stories strung in a silky web, are held up to the bright stars which reveal some areas of light and dark, room for growth and metamorphosis. With this in mind, I have separated the methodology of ecoportraiture into two stages: 1) Collecting and gathering 2) Building the case and holding these stories up to the stars.

✖

Despite being a lifelong "nature-lover" and environmental educator, I admit, I had never heard of caddisflies before the students started pointing them out and Laura explained what they were. I had seen them before, many times—I also spent my childhood wading into coastal salmon-bearing rivers and turning over rocks—I just never recognized those crunchy "river bugs" as worthy of deeper consideration. I join the group huddled around the caddisfly and listen to Laura weave "empirical observations" with metaphors and stories, drawing the students into a world where larval "underwater architects" construct homes made from silk and alluvial debris, before casting it all away and swimming up towards the stars to take flight. I nod knowingly—waiting until all the students have had a look and started down the trail—before sneaking a few moments of close-up "participant observation" myself. How had I missed this? Where have I been? I carefully set the rock back into the riverbed, scooping water into both hands and splashing it onto my face, sobering-up, as it were, from a wonder-induced trance and double-timing it down the trail before I miss something else. Mental note: pay attention, pay attention, pay attention—like a child in a river.

✖

Metaphor matters. For instance, I teach action research courses for in-service teachers, which invariably include a historical overview of qualitative methodologies in education. It can be a jargon-heavy undertaking, particularly for those new to thinking about knowledge production in terms of epistemic location. The concept of *bricolage,* in particular, always seems to prompt bouts of academe-weary groaning at first. Explaining the term definitively as a "multiperspectival research method" or a "mode of multimethodological inquiry" does not help matters much. But when we discuss bricolage as a metaphor for making sense of the research process itself *as* quilt-making, or mixed media collage, or assembling texts into a film-like montage, something happens. Most teacher-researchers discover they were bricoleurs all along and that the metaphor has helped facilitate a new means to traverse their own mountains of data. They tend to feel free to wander a little, beyond the pale of positivistic ordinance; free to collect and gather bits of stories here and there, careful not to do violence, careful not to expect too much, careful—as I implore them again and again—to take their cue from that which is given as to how the gathering must go. This is, I believe, the true spirit of bricolage: a way to "think through" the lived experience of conducting research in impossibly complex settings. The narrative turn in the social sciences over the last few decades, impelled by and including portraiture, has provided the free space to traverse epistemic terrains that, in the past, have been methodically severed from what constitutes "valid knowledge." And as the positivist hangover continues to wane, the real discovery is the one that will allow us to *ecologize* the narrative turn and resume thinking metaphorically when we need to—i.e. when it is called for(th) by the world.[3] This is not, I think, exactly what Lawrence-Lightfoot had in mind when she described portraiture

[3] This insight is informed by the lyric philosophy of Jan Zwicky (1992, 2014). Also see Robert Bringhurst's *The tree of meaning* (2006) for more on the relationship between nature, poetry and ontology. I have explored the significance of lyric philosophy, metaphor and environmental ethics in education in a book titled *Place, being, resonance* (Derby, 2015).

as blurring the boundaries between aesthetics and empiricism, but if we take "empiricism" here to mean the kind of disciplined attentivity required by ecologists and nature poets and artists "in the field"—I think the true spirit of portraiture remains both intact and revitalized.

✖

"So ecoportraiture is qualitative research with a bunch of metaphors then? Like, lyric inquiry or something?" Perhaps, but there is, I think, something of an "ontological turn" required to conduct good ecoportraiture that takes it beyond methodologies that are simply rich in metaphor or other poetic devices. We are not, to play out the motif, merely transitioning from portrait to landscape painting (education invariably involves humans and, notwithstanding "direct experience," language-using discourse communities), rather we are attempting a different mode of composition that attempts to capture the aesthetic whole of human experience as nested within a "natural world" that is every bit as ontologically robust. A more-than-human gestalt, as it were. Art works co-created as multispecies collaborations might be the closest analog, but the metaphor does not translate all that well to the page. I am tempted to suggest the surreal portraits of 16th century Italian artist Giuseppe Arcimboldo—who painted composite figures composed of various fruits, vegetables, flowers, fish and books—these portraits could, however, still be construed as fundamentally anthropocentric. [That being said, there is something undeniably fruitful here and I think research that explored the more-than-human co-constitutionality of our bodies and minds in educational settings would be a fascinating avenue for ecoportraiture—see Affifi, 2016]. But if we can move beyond visual analogs and switch senses, I believe the best metaphor to think through ecoportraiture would be that of artistic soundscape ecology; that is, polyphonic compositions informed by attending thoughtfully and seriously to both human and other-than-human "voice" as aesthetic whole. Textual compositions of ecoportraiture would thus be aimed at capturing diverse human experiences (i.e. the multiperspectivalism of conventional portraiture) without drowning out the multispecies polyphony of the "underlying" natural world. The tendency of what soundscape ecologists refer to as the *anthropophony*—i.e. human-borne sound—to drown out the *biophony* and the *geophony*—sound produced by non-human lifeforms and earthen processes respectively—offers an obvious scientific analog. No easy task. But if we are to "go eco" with portraiture, perhaps the "search for goodness" becomes not only a commitment to not pathologizing human children, but also refusing to objectify other non-human subjects. Goodness, in this sense, might describe the trace of a kind of epistemic reciprocity with the other-than-human world. Something closer perhaps to how Zwicky qualifies "good" poetry: "Good poetry, like all meaningful thought, traces a gesture of address. It enacts ontological attention." Good ecoportraiture, it seems to me, also ought to be informed by, enact and provoke such ontological attention. As Zwicky claims, "Metaphor is one of the means [poetry] uses to do this."

✖

At the risk of oversimplifying an immensely complex topic, let's begin with the notion that metaphors are *cognitive tools*—they are *bon à penser*; that is, good for thinking with (Egan, 1997). But, and here we come to the ontological turn, does Laura employ the caddisfly metaphorically as a way to "think through" her methodology or did she actually "dialogue" with particular caddisflies as co-researchers? When I put the question to her she seems understandably hesitant to claim the ability to converse with caddisflies and suggests it's more like *seeing* differently via focused analogical thinking. But crucially, she did not just "make it up" in her head, she listened, seriously—careful not to expect too much, prepared to wait, prepared—dare we admit it?—for the possibility that nothing would come forth. This is the beautiful risk. In other words, over the course of her research project Laura cultivated a lived practice of *being with* "wild others," informed by eco-ethical premises, to facilitate a reciprocal way of knowing such that her thinking resonated *with* place (place being manifest, in this instance, as a caddisfly in the Alouette River). "By stepping back from the need to construct a story from the outset and listening rather to what emerges, I felt that I had been learning how to co-create the story *with* place and the students" (Piersol, 2015, p. 46). At the end of the day, what is the difference between going for a walk in the forest and doing ecological field work as a researcher? On some level, it's simply a matter of paying (a different kind of) attention. This is why metaphor matters epistemologically *and* ontologically—a good metaphor says: "Well, have you looked at it like this?" A good metaphor can also be a signpost pointing to the world and, suddenly, one is *struck*: How have I missed this? Where have I been?[4]

[4] Jardine explains Hans Georg Gadamer's hermeneutics in terms of a similar initiatory phenomenon or aesthetic experience. "Before we adopt any methodological stance, before we 'do' any research, before we know it, we enter a classroom (sic) and something that a student says, some work they have produced, something they have written, some question they ask, the look in one child's eyes, some sketch posted on a bulletin board—these simple things sometimes *strike* us, *catch* our fancy, *address* us, *speak* to us, *call* for a response, *elicit* or *provoke* something in us, *hit* us, *bowl*

✖

Perhaps metaphors are not *only* cognitive tools then? Not only thinky things floating about in human grey matter and expressed in human language? Sometimes perhaps, after many kilometers on a mountain trail or many hours of contemplation beside a river, they can be a means of tracing an address? Indeed, when Laura tackles the ever-contentious issue of anthropomorphizing non-humans, she suggests approaching such perspective-taking with the kind of "poetic attention" described by Don McKay. "Poetic attention is based on a recognition and a valuing of the other's wilderness; it leads to a work which is not the vestige of the other, but a translation of it… Poetry should not be taken to be avoiding anthropocentrism but to be enacting it thoughtfully" (1995, pp. 25–26). This by no means *solves* some of the more philosophically problematic aspects of multispecies representation, and I recognize that teacher-like demands to *pay attention* are not the most satisfying or concrete of methodological instruction—but truly the fundamental gesture of this work is deep, immersive listening.[5]

us over, *stop* us in our tracks, makes us *catch our breath*" (2006, p, 270). In their book, *Conducting hermeneutic research* (2015), Nancy Moules et al. also describe this phenomenon in similar terms as the "address of the topic." While both works tend to discuss the topic in terms of human activity and artefacts, there is no reason why "the address" needs to be confined to the human (for ecological readings of hermeneutics see: Clingerman et al., 2013; Derby, 2015; Jardine, 1998, 2012; Siedel & Jardine, 2014).

[5] There is an important distinction to be made here, particularly in the context of conducting research within settler colonial states, between allowing our thinking to be metaphoric in the sense of resonating with place and employing ecological metaphors to shape thinking without reference, prolonged experience, or dialogic relation with actual places. The latter tends to be more common amongst academics and researchers given the marginalization of place in the social sciences and it is important that ecoportraitists avoid inadvertently replicating epistemic violence against Indigenous ways of knowing (Smith, 2012) or denigrating the dignity of other-than-humans by "translating" their existence into metaphors *for* human thinking. This is, in fact, the very opposite of McKay's notion of poetically attending to place. The term "ecosystem" may be a handy way to describe the internet and bees may provide an apt symbol for "hardworking," but this does not mean there is anything inherently "ecological" about the internet, nor should it detract from the intrinsic value of *this* particular bee. Indigenous and place-sensitive scholars begin to get justifiably concerned not only when place is flattened to an inanimate background surface upon which data is collected, but when the discourses employing the language of ecology detach from the particularities of place and place-based social practices altogether. On the one hand, ecoportraiture, like all place-based research, inherently runs a high risk of reproducing colonial logics, which is why it is so crucial that ecoportraitists are informed by a kind of *critical* place inquiry (Tuck & McKenzie, 2015).

✖

Ultimately, ecoportraiture requires practice *in the field*—immersion and time and discipline, like learning a new language abroad or how to traverse difficult mountain routes—in order to get better at "getting it." I submit we could never hope to capture the essence of a place like so many panoramic photographs or landscape paintings; the interconnected, living gestalt of human and other-than-human beings resists that kind of straightforward representation. But realistically, could we ever hope to capture a standardized essence of a classroom with its somewhat intelligible human behavior and utterances either? Such work is always risky. The search for goodness in ecoportraiture is thus an honest attempt to foreground what has been historically backgrounded and pathologized in social science research (read: the dominant culture): thoughtfully enacting the voices and, where possible, perspectives of the other(-than-humans), thinking metaphorically when we need to, cultivating lived practices and reciprocal ways of knowing by *being with* wild others. "So part of your research methodology is just being out on the land and… listening!?!" Yes.

✖

> How do we come to entertain doubts about some deep-rooted cultural orthodoxy? By checking beliefs our culture invites us to hold against a background we hold to be even more secure: our direct experience—what happens when we pay attention to human beings of other races or genders, what happens when we pay attention to nonhuman beings, what happens when we pay attention to what we're seeing through a telescope? But can't direct experience be coloured by racist, sexist, anthropocentric, or terracentric prejudice? Yes. And so we must struggle to be honest with ourselves. Paying attention is hard work. It takes courage—in some cases, to see what's there, in some cases to admit we've been wrong. It takes time. It takes patience. (Zwicky, 2019, p. 53)

✖

From *Eco care: Nurturing possibility & resistance within education*:

> During my research the students and I would often intentionally pose questions to each other to consider the world from the perspective of an individual

> On the other hand, if the fundamental gesture of ecoportraiture entails *being with* and listening other-than-humans and attending to place with poetic sensitivity, this inherently problematizes settler relations to land and, one hopes, begins to nudge research in the direction of alliance with both Indigenous scholars and other-than-human beings.

> creature in the natural world... I asked one of the students, Raven, to take me to the fort that she had been building in the woods, and while we were there I offered her a camera to take photos of it... I decided to try an experiment; this student has chosen the 'pseudonym' Raven because she feels connected to the birds so I asked her to photograph the fort from a raven's perspective:
>
> I am struck by how she actually begins to move like a raven. She tries multiple different angles, swoops down and above. She is not concerned about documenting certain parts of the fort anymore but instead enters spaces where she thinks the raven would perch. She holds the camera for longer and takes more pictures. In my sit down interviews with Raven at the beginning of the year, she talked slowly calmly and quietly, listening to the audio one can hear the river and a raven in the background as we are talking but those voices remain there, in the background. In this approach I can see Raven open up and talk faster. Her eyes get wide and she begins to skip, usually ahead of me, excited to show me around. She eagerly asks me if she can use the camera. I see a whole different side of her when I allow the methodology to be more creative. We also experience a different way of being in place, where Raven, the bird, is perched on the forefront of our thoughts rather than separated in the distance.
>
> In this way we were playing with learning about the raven's way of being in the world, what it sees and experiences, and what we can learn from it. "I like thinking of the world like it would, I know I can't understand what it actually sees but I feel like I understand it more when I try." Raven commented (Piersol, 2015, pp. 29–32).

✖✖✖

II. *Autoethnography-in-the-Forest as Gesture Drawing:*

A First Sketch

Disclaimer: The following section is *not* an example of what I would consider "good" ecoportraiture; in the sense that it mostly fails to evoke the voice, agency and intrinsic value of other-than-humans. Both the form and content are lacking in polyphonic resonance, as it were. Rather it's something closer to *critical* portraiture with an ecological emphasis, or autoethnography-in-the-forest as I have called it. The account is based on the collective field notes of several place-based researchers at the Maple Ridge school project (despite the fact it is written in first person) and represents an initial attempt at taking portraiture in a more ecological direction. Jessica Hoffman Davis has described such initial sketches as *gesture drawings*—brief narratives scribbled quickly to capture meaning, as opposed to carefully crafted pieces—as a means of refining portraiture skills for those new to the methodology

(Lawrence-Lightfoot & Davis, 1997, pp. 22–23). There is, I believe, much value in a negative example and I will offer some reflective thoughts on this gesture drawing and on my research process generally for prospective ecoportraitists in a following section (IV.).

✖

> Portraits are always framed by the ecological context: a vivid description of the geography, the demography, the neighbourhood, and a detailed documentation of the physical characteristics of the place that evokes all of the senses—visual, auditory and tactile. The reader should feel as if he or she is *there:* seeing the colours of the autumn leaves, feeling the temperature of the gentle breeze, and hearing the rustling of branches. That is, he or she should not only see the contours and dimensions of the terrain but also feel placed in it, transported into the setting. (Lawrence-Lightfoot & Davis, 1997, pp. 44–45)

Heavy rainfall begins seeping into my dreams ten minutes before I am consciously aware of what it means. From the warmth of my sleeping bag, the cadence is hypnotic, sleepy—but relentless alarm bells finally force me to acknowledge the brisk realities of late November in the Coastal Range. It is still dark and the day is already completely waterlogged. Despite being a "place-based researcher," I do not necessarily want to go outside today. I am reminded of one of John Locke's opening thoughts concerning education: "The first thing to be taken care of, is, that children be not too warmly clad or cover'd, winter or summer. The face when we are born, is no less tender than any other part of the body. 'Tis use alone hardens it, and makes it more able to endure the cold" (1693/1996, p. 11). I question both the physiological and pedagogical logic of this statement, but then again, there is some kernel of truth here immediately recognizable to any outdoor educator. One of the most common remarks I hear as a field researcher at the Maple Ridge Environmental School is how accustomed the students are to being outside in the rain. "They are building resilience," say teachers and parents with great pride. And I admit, I too have been surprised by how quickly students have acclimatized to such unconventional conditions—but is this what education is for? To toughen up to the elements?

Initiate morning ritual. Eat breakfast, tidy up, and begin layering on the warm clothing. I am without a packed lunch, again, because I stayed up reading about local edible plants and Lockean pedagogy. On the way to Malcolm Knapp Research Forest, I stop for a sandwich and coffee-to-go and notice a newspaper article on the Fibonacci sequence in nature. The article has a picture of a large pinecone painted with two whorled lines up its sides to illustrate the golden ratio manifest in the arrangement of its scales. This would make a great math project for some of the older students I think, as I turn my back to the barista and pretend to cough while ripping out the page. Time to get to "school."

✖

> The alternative that portraiture provides raises a reflective glass to the stories that shape lives, pedagogy, and institutions. In doing so, portraiture illuminates and acknowledges the importance of these phenomena. The close observation that portrait writing requires unites the researcher to the subject—like the artist assuming the pose of the model—and affords a view of the parts of the whole that insiders know intimately. (Lawrence-Lightfoot & Davis, 1997, p. 36)

The parking lot at Malcolm Knapp is dark and sheltered by at least a dozen massive Western Red Cedars and a few smaller Douglas Firs—all marked for death with blue and yellow spray paint. I wonder if there has been any discussion at the school about the loss of these giants to make way for new power lines? I have heard a few comments regarding safety imperatives, others shrug and surrender to the inevitable—we all need power, right? I sigh, lamenting my consistent failure as an educator to delve into more complex political and ecological topics with young people in what I would consider *empowering* ways. I used to teach high-school aged youth-at-risk, but I left that to pursue research here because I wanted to work further "up-stream" with elementary-aged children, before... the "inevitabilities" got hold of them. After three years of research, I am not sure there is an "up-stream." There are no pristine headwaters up in the mountains somewhere, above all this culture and commerce, from which we spring forth with glacial purity. We are all born downstream of the megaproject that is modernity—and school is the dam that powers the whole enterprise. I imagine what I might say to students about the trees, how I might respond to the question: "why are those trees spray-painted?" I know the answer that keeps us all on course: we all need power, right? I wonder about what it means to use the schooling system to cultivate doubts about deeply-rooted cultural orthodoxies? Is it even possible? Would it make a difference if we talked about the trees inside the classroom versus marching over here, touching these particular beings, listening for a moment, gathering our thoughts *with* them? How might the trees contribute if given the dignity and recognition to do so? But everyone knows, trees cannot speak.

✖

> In portraiture, then, the place and stance of the researcher are made visible and audible, written in as part of the story. The portraitist is clear: from where I sit, this is what I see; these are the perspectives and biases I bring; this is the scene I select; this is how people seem to be responding to my presence. (Lawrence-Lightfoot & Davis, 1997, p. 50)

Time to get moving, I steal one last guilty moment of petroleum-generated heat in the parking lot before exiting my vehicle. As I pass by the welcome sign I realize I have forgotten for the thousandth time to research just who Malcolm Knapp *really* was. I have also been meaning to look up Simon Fraser. I know the former is a celebrated "forester" and the latter an esteemed "explorer," but these official epithets have the suspicious redolence of colonial revisionism. This place bears their names, but neither man was *of* this place (both were born in what is now New York State—I finally looked it up).[6] The "Fraser River" is not *really* the name of that river, even though that is the story we habitually teach here. There are, of course, other stories; many could be told. One might tell a tale of young Simon, the aspiring corporate protégé, whose imperial exploits in this area paved the way for the catastrophic despoilment of the river that now bears his name. A mountain and a university honor him, but his legacy, and by extension, "our" legacy (all colonial-settlers to come after him—most of the students here, my own family, myself), is also defined by the pulp mills and farmland run-off, the collapse of the salmon run, the near extinction of white sturgeon, not to mention the devastating loss of a way of life sacred to the Katzie, Kwantlen and all other *Stó:lō* peoples (the peoples *of* the river). And yet every time I look to my degree from Simon Fraser University I see his name arched above my own and every time I refer to the river I say "Fraser" instead of using, or at least supplementing with, the Halq'eméylem name. I want this school to be the beginning of a new legacy, but it's a struggle to even get the names *right*.

I take one last sideways glance at the information board, which also features a large photo and reverent commentary on one of the antiquated "Empire" steam donkeys used to log this area in the past. The younger students often climb on the winches and gears or try to shimmy up the boiler of the old behemoth permanently parked at the border of the "play area." I consider how so many of us—teachers, researchers, parents, superintendents—have worked so hard to "get outside," to get "hands-on," to get back to "the real world." But even here natures and cultures clash and intermingle

[6] Apparently there is a historic marker in Mapletown, NY – the birthplace of Simon Fraser – that endows Fraser with the title: "Explorer of Western Canada." The plaque adds: "A river and mountain there honor his name."

and the young invariably play on the structures of the old. There are so many stories that might be told.

✖

> In this illustration, we see the portraitist at work, notebook in hand, pen poised. But we also look inside her head and heart and see the ambivalence, the contradictions, and the turmoil. (Lawrence-Lightfoot & Davis, 1997, p. 52)

I see morning attendance is finished and students are heading to and from the yurt for reading time. I strategically find a camp chair near the fire in order to "conduct some research" (read: warm-up). Younger students swarm me with books in their hands pleading to read to them. I acquiesce—noting that the books these students have access to are so much more "environmentally friendly" than the books I used to read in elementary about Tonka Trucks and "Indians" and Candyland. One young boy *loves* reading about insects and has become the resident expert on bug identification. I wonder if the same interest would have formed in a conventional classroom with conventional resources?

Sometimes I catch kindergarten kids staring quizzically at me while I am reading—"What are you doing?" they ask incredulously. "I'm reading." Some continue to stare, half smiling, trying to determine whether or not I am trying to trick them somehow. I think they are trying to understand why I would sit here quietly, alone when I could be running around the forest playing? Others look to the dense reticulum of text, considering whether there might actually be something in there worth knowing, worth traversing the long, arduous path of literacy for. I love books. I wish I could convey to these children all the wondrous gifts that books have given me. I was fortunate to have a father who took me into the backcountry and a family life that centered around camping, but my "ecological consciousness" did not really flourish until I first encountered nature writers and philosophers. Learning to read, however, does not come naturally for everyone.

As I look around the yurt and outside in the rain, huddled beneath shelters, I see many students absorbed in their books or sharing stories in small groups. As someone whose school experience was dominated by rowed desks, comprehension worksheets, and required readings, I envy these students a little, nestled as they are by a fire or hunched beneath a tree, with the freedom to read any book they choose. As a teacher, however, I cringe a little at the prospect of tracking so many readers at so many levels in such an *organic* environment. I cannot help but notice how many students are "off-task," climbing trees and playing games. Perhaps I am too entrenched in my own presuppositions about what education is and what it is for? Too much a product of my own instrumentalist schooling and teacher training? On the one hand, there are compelling arguments that suggest becoming literate may entail a loss of immediacy and participatory reciprocity with the world (Abram, 1996). If this is so, one of the objectives of environmental education ought to be mitigating such losses, perhaps by delaying the acquisition of literacy and focusing on more play and direct contact with "nature." On the other hand, I shudder at the thought of graduates who are "nature-loving," but functionally illiterate. And what of my own experience of a relationship with the world deepened through literature and philosophy? The research team has high hopes that this school might cultivate a kind of deep "ecological literacy," not only learning *about* the principles of ecosystem organization and sustainability typical of eco-literacy programs, but learning *from* and *with* this place. We have managed to immerse reading (and researching) *in* the natural world, and for the most part *about* it, but how to read and research *with* this place?

✖

> The portraitist's stance is one of acceptance and discernment, generosity and challenge, encouraging the actors in the expression of their strengths, competencies, and insights. She sees the actors as knowledge bearers, as rich resources, as the best authorities on their own experience. She is interested in examining the roots of their knowledge, the character and quality of their experiences, and the range of their perspectives. (Lawrence-Lightfoot & Davis, 1997, p. 141)

Nature calls. I begin trekking across the field to the resident outhouses, nimbly dodging the puddles and waterways that swell and merge as the rain continues to pour. Passing by a gaggle of Kindergarten girls clad in bright rain gear I hear one of them say: "Now add the peppermint salt." Peppermint salt? What poetic genius is this? I cautiously approach the group gathered around a puddle and inquire as to the origins of this minty crystalline magic? The secretive coterie immediately falls silent. "Come on," I implore, "what *is* peppermint salt?" They giggle and look into the puddle, one picks up her stick and begins stirring the elixir again, mixing in secret ingredients I will never know. I reluctantly leave the little conjurers and prepare myself for the unpredictable state of... the facilities. I imagine the stark reality of outhouses must exclude a good percentage of teachers, substitute teachers and students from attending this school from the get-go. I laugh to myself on the throne, considering the absurdity of "modern" humans; how quickly we become accustomed to certain comforts and amenities. Central heating, cell phones, smart boards, global positioning systems, jungle gyms, central plumbing.

And likewise, how quickly the young reacclimatize to new settings and realities—cold rain, muddy boots, an utter lack of rowed desks. This gives me a small whiff of hope in an otherwise miasmic situation (outhouses, incidentally, are ideal environments to contemplate existential conundrums).

✖

> Reaching beyond the immediate experience of subject or site for contextual elucidation is an important step in informing context. Although a lens is being focused on a particular setting and stage, the context determines the curvature of the lens and the illumination of the view it provides. The result of far-reaching attention to multiple sources of data is a collage of carefully chosen facts, views, voices and impressions. (Lawrence-Lightfoot, 1997, p. 64)

As I head back toward the relative warmth of the yurt, my teacher safety sense starts tingling and I notice two Grade Three and two Grade Four boys climbing rather high in a deciduous tree set back amongst a stand of conifers. As I approach, I see that they are engaged in a "moss ball" fight, rapidly denuding the tree of the creeping feather moss that covers it, balling it up like snow and throwing it at each other. We have discussed "our impact" on this place many, many times. I take a deep breath and prepare myself for role of the heavy. On the one hand, one might reasonably claim: it's only moss. This response is certainly supported by the dominant culture which tends to view moss as scientifically alive perhaps, but certainly not sentient, not valuable, not important like us computer-wielding humans; and not, generally speaking, worth reprimanding students for. On the other hand, we are an "environmental school" and one of our objectives is to exert as little impact on the places we visit as possible. The principal has consistently reiterated the importance of respect as an action—respect is something we do. On the other, other hand, one might reasonably claim that moss is living in ways that may extend beyond our scientific definitions; it's embedded in a complex web of life with its own inherent dignity, sensations and right to exist. For me, it's this depth and mystery of ecological understandings, of coming to know our "implicatedness in life" (Orr, 2005, p. 105), that lies at the heart of this project.

I approach the boys (who stop immediately and avert their eyes as soon as they see me) and despite my best intentions to turn this into a "teachable moment," a pejorative mood sets in. "Gentlemen, is that moss alive, do you think?" I ask. Half shrugs and downward gazes, silence. "I think we've talked about this, right...about respecting this place? Is this what respect looks like?" One offers a half-hearted "no," the others refuse to look me in the eye and begin climbing down the tree. I walk away half-hearted myself, not exactly the profound post-anthropocentric intervention I was hoping for. I cannot help but feel, from their perspective, the moss is just dumb moss, it's everywhere and balling it up is fun, and I am just another overbearing adult ruining their play. How, I wonder again, do we draw their attention, as distracted as it is by modern (de)vices, and entrenched as it is in cultural presuppositions, towards a deeper sense of our implicatedness in life? I stand in the rain and watch the boys from the other side of the field, curious as to whether anything took root. They linger around the tree for a minute and then run wildly across the field to join some other students splashing in puddles.

I consider the tree with dying moss balls scattered around the base of its trunk for a moment; my mind's eye zooms out and hovers fifty meters above the field and small wooded area around the yurt where the children play games and learn wonderful things; I zoom out above Maple Ridge, taking note of the ever-expanding developments reaching towards the borders of the research forest; I zoom out above the southern part of British Columbia (an image supplemented by a recent flight from Toronto to Vancouver) and lament the vast clearcuts beyond the beauty strips, the disturbing patchwork that was once interconnected forest. I zoom out and out into the cosmos where "nothing matters" and just... float in orbit for a while around a distant star. But somewhere deep inside I know I'm not destined for the vacuum of heaven; I live here, amongst *these* trees and *that* steam donkey and *those* children and *this* place. I turn back towards the yurt with a fistful of dying moss in my hands and a crazy notion. What if we did not teach *about* moss, but *with* it as a kind of teacher in its own right?

✖✖✖

III. From Eco Care: Nurturing Possibility & Resistance Within Education

> The moss unit... had an interesting effect on one of the older students... so we sat down and chatted. I asked her what she had learned about moss. "Laura, it's crazy" she tells me, "I think of it almost like a person!" I ask her to elaborate "I would have never said this before, you know me.... [her family is strongly into hunting and she has long been vocal about how she sees the land as a resource]. But, the more I learned about it, I felt like I was learning about its 'culture,' like where it came from, what else it was related to..." Curious, I asked "is this about moss in general or the particular sample of moss that you were doing your inquiry on?" She answers, "It's the specific moss that I was studying." She tells me this with big eyes as if she's confessing something ridiculous: "I felt connected to it after all those lessons. I even wonder why we had to take a sample. There are lots of us as kids, surely that hurts the moss. We could've observed

without taking. So I went and said 'thank you for this learning opportunity.'" She laughs, "Do you think I'm crazy?" (Piersol, 2015, p. 200).

***

IV. *Risks & Rewards of Traversing Free Spaces: Some Thoughts on Doing Ecoportraiture*

As Lawrence-Lightfoot notes, very few social science researchers have "tried to seriously describe the process of creating a gestalt"; in part, one assumes, because aesthetic wholes tend to resist the reductionist modes of exposition required by most methodology texts. Art critique is rife with the kind of tensions that emerge when the analytic mind confronts an artefactual trace of gestalt comprehension. "What were you thinking when you created x?" The artist looks back on her piece, "Uh… nothing." Process tends to be "difficult to explain" and sometimes one gets the sense that the commentary generated by analysis is more retroactive projection than any kind of original intent. Be that as it may, I would like to attempt some reflection on process by way of critiquing my gesture drawing and providing a few suggestions for aspiring ecoportraitists. I will conclude by reiterating the importance of risk-taking in this work, something the gesture drawing here does a reasonably good job at exemplifying, with a focus on upholding (and ecologizing) the portraiture tradition of evoking both the critical and the beautiful.

The problem with the gesture drawing above, as I alluded to in the disclaimer, is that it's not very "good" (i.e. not very *eco*); in the sense that it does a poor job enacting the kind of ontological attentivity I claimed was prerequisite in the first section. Do not get me wrong, it paints a fairly good picture of daily life at the Maple Ridge Environmental School during that time; that is, it's a decent example of eco-friendly portraiture (i.e. autoethnography-in-the-forest). But it lacks the thoughtful interweaving and foregrounding of the (more-than-human-) other that one finds in Piersol's work for example, and that ought to be more palpable in good ecoportraiture. In terms of our soundscape ecology metaphor: the anthropophony is simply too loud (as usual) and the writing mostly drowns out any sonorous sense of a multivocal, multispecies polyphony. I could make excuses: it was a collective writing experiment; it was written before the research team started discussing ecoportraiture as a distinct methodology; it's time-consuming and culturally taboo to build relationships with caddisflies and creeping feather moss as a school-dumbed settler, etc. But let's just acknowledge it for what it is—a first sketch—and focus on lessons learned.

No two ways about it, engaging in an "ontological turn" can be a disorienting, alienating and often painful process; particularly with respect to recognizing other-than-human subjectivities and coming to grips with the way the dominant culture—and by extension our own lifestyle—objectifies and exploits living beings as "natural resources" (Blenkinsop, 2012). Most folk do not seriously entertain the notion of multiple ways of knowing and, indeed, much of our material and institutional culture is aimed at reinforcing extractatvist and colonial norms (conventional schooling, for example). I suppose it's not appropriate to make eco-ethical crisis or generalized social estrangement a requirement for conducting research, but given the diminished state of things—biodiversity, democratic institutions, the attentive capacity of device-wielding humans—it's a very real possibility in this work that ought to be embraced *as* research. At various points throughout the project, I think every member of the Maple Ridge research team went through something of an existential "rough patch," which I suspect is ultimately why portraiture appealed to so many as a guiding methodology. Portraiture, despite the potentially misleading metaphor, is very much a reciprocal approach that allows one to explore the dialectic transformation of both research subjects and one's self. The researcher's reflexive process of becoming is, or at least can be, valuable research. Moreover, if the search for goodness in ecoportraiture is, as I suggest, on some level a search for deeper ethical relationality and reciprocity with the other-than-human world—this dizzying up and down, emotion-laden, revelatory process ought to be a core component of the work. Exploring the clash of different ontological orientations—the moss incident above, for example—makes for fascinating research about how our underlying presuppositions about the natural world are played out, reinforced, and, one hopes, challenged in educational settings.

While the researcher in my gesture drawing is highly reflexive about many of the pedagogical conundrums one encounters, there is little commentary on the phenomenology of heightened eco-sensitivity or its origins other than a few vague references to reading "ecophilosophy" books. In other words, the sketch does a reasonably good job at capturing the voice of the researcher *in* place and broadly concerned *about* place-based topics, but one wonders (yet again) what it would mean to conduct research *with* place? To take our cues from that which is given as to how the gathering must go. To conduct and compose research where, to borrow from one of the principles of the Maple Ridge school, "we listen for what the more-than-human world has to teach us." This is, I believe, the measure of good ecoportraiture, and ultimately the challenge that future researchers must take up. My gesture drawing, to provide a negative example, lacks what we might call deep ecological reflexivity; in part, I think, because I was not as serious or explicit about my "eco-ethical premises," as Piersol was in her work. This results in a sketch that is, if I may be so bold, refreshingly candid and colored by ecojustice concerns, but lacking ontological depth in terms of formal innovations, philosophical inquiries, or writing that "points to" or "makes space" for other-than-human voice(s). One requires serious consideration of eco-ethical premises as guiding lights in order to first perceive "beyond the human," and then to inform the story-weaving process—fraught as it is with the crisis of representation. This is, I think, what Piersol had in mind when she poetically described her process as twofold: 1) collecting and gathering, and 2) building the case and holding these stories up to "the stars" (i.e. the guiding lights of her premises). Note: this is *not* to say that researchers are required to confidently state their ontological and ethical presuppositions up-front, just the opposite as I will discuss below, but that making such process-based inquiries, conundrums, and premises explicit is part of a commitment to courageous conversations that will, one hopes, give the portrait more ecological depth and make for more evocative research. As Zwicky has suggested, attending to the phenomenology of experiences that are marginalized by ideological emphases in the dominant culture is one way to "begin to reflect on the politics of their marginalization" (2014, p. L109).

In my action research courses, I have started suggesting that teacher-researchers consider integrating their theoretical framework and literature review sections into a reflexive narrative as part of an evolving "research journey" that will extend the length of the project. I find this a more honest approach as most educational researchers are, for better or worse, simultaneously reviewing the literature, being exposed to new theories and ways of knowing, developing epistemic reflexivity, and conducting field work all at once (and many are in-service teachers on top). Ideally, this frees researchers to wander a little with bricoleur-like abandon, both methodologically and philosophically, picking up pieces and practices that ring true without pigeonholing themselves too tightly in a research or intellectual tradition. The kind of reflexivity and attentivity required to do ecoportraiture takes *time* to develop—and there will likely be "rough patches" along the way—so why not be honest about the pitfalls, cognitive dissonance, and insights as they emerge? Not only does this help teacher-researchers traverse some of the potential disorientation and doubt they may experience in the face of ontological readjustment, it tends to make for more evocative, relatable, and thought-provoking research for the reader. I suggest this research journey narrative include a contextualization of one's initial presuppositions about the "natural world," perhaps by providing some locational background or experiences from one's own schooling or childhood, and then charting changes and inquiries as one engages with the literature, participates in discourse communities with "spiritual comrades," and/or deepens their own, for lack of a better term, "nature practice(s)." This ought to go without saying, but this implies, of course, that part of the process of conducting ecoportraiture is reading a grad-school amount (i.e. a shitload) of books and articles that provoke ontological reflexivity.[7] And, perhaps less common in grad school, but equally as important, coupling this literature with immersive experiences and contemplative practices "in the field."

After such a thorough ontological thrashing of my own gesture drawing, I hope you will indulge a few final comments about some of the things I think this sketch does well—namely voice as a research instrument and an uncompromising critical consciousness—which, I think, ought to be hallmarks of any future ecoportraiture. For me, the two are inseparably entwined and foundational features of the portraiture tradition from the very beginning. Presumably it's possible to compose portraits that are rich and evocative with respect to researcher voice—exceptionally well-written accounts with compelling "characters" and insights into the human condition, etc.—but otherwise lacking a critical dimension. These would, however, be exceptions in the tradition and on some level "not good" or appropriate in a time of ecological emergency or in the context of settler-colonialism. Lawrence-Lightfoot has stated from the beginning that portraiture is aimed at *seducing* readers into thinking more deeply. And as Mark Fettes, Sean Blenkinsop, and

[7] There are numerous (and often incommensurate) philosophical positions one might employ to recognize the significance of place and other-than-human beings in research. Piersol, for example, draws primarily on the work of Jacques Ranciere (2004), Martin Buber (1947), and Val Plumwood (1993, 2002), in addition to a host of contemporary ecophenomenologists (Abram, 1996) and ecofeminist scholars. Others might look to current trends in multispecies ethnography (Haraway, 2016), the non-human turn (Grusin, 2015), sociomaterial practices (McKenzie & Bieler, 2015), philosophical animism (Bai, 2009; Barrett, 2013; Rose, 2013), or the "ontological turn" in anthropology (Descola, 2013; Kohn, 2013; Latour, 2013; Viveiros de Castro, 2014). Not to mention looking to the scholarship and research methodologies of Indigenous scholars in their own words (Smith, 2012; Tuck & McKenzie, 2015). In my own work, I have tended to employ the hermeneutic tradition to make sense of language, metaphor and meaning in a more-than-human world (Derby, 2015). Two recent publications that might be helpful are *Interpreting nature: The emerging field of environmental hermeneutics* (Clingerman, Treanor, Drenthen, & Utsler, 2014) and *Carnal hermeneutics* (Kearney & Treanor, 2015).

Laura Piersol illustrate in the introduction to this collection: "one cannot fully make sense of the portraiture tradition without acknowledging its connections with a long emancipatory tradition in African-American culture, or with its fundamental orientation towards social justice and a compassionate feminist axiology" (see Chapter 1).

I cannot speak as to whether my gesture drawing was successful in seducing readers to greater critical awareness, but in terms of original intent, I can attest to the fact that traversing the mountain for me was a process of drawing out the instances I thought might provoke the most critical awareness (e.g. the discourse analysis of the info board at Malcolm Knapp forest, for example). Many stories might be told—for me, the ones that draw out critical consciousness are primary. Fettes, Blenkinsop, and Piersol also suggest in the introduction that (eco)portraiture is essentially a hermeneutic mode of inquiry; an insight I wholeheartedly agree with and, I think, is illustrated in the necessary dialectic between the critical and the beautiful in this work. In other words, hermeneutic approaches have a critical dimension, but they are not the same as research inspired by critical theory. And two of the principle differences for me, both foundational to portraiture, are the reliance upon voice of the self as an instrument of inquiry and the relevance of the beautiful to achieving verisimilitude. While much has changed in research culture since the publication of *The Art and Science of Portraiture* (1997), the risk entailed in conducting this kind of research is still very much real—particularly if one "goes eco." Cultivating authentic voice puts one in a vulnerable position and can lead eco-sensitive researchers to wonder, much like the older student after the moss unit, if they sound "crazy?" My approach to this matter is to embrace radical vulnerability and to try to establish a relatable subjectivity with the reader as a core objective of the writing early—in addition to, or ideally before, barraging them with epistemic doubts about their core beliefs. I tried to achieve this in the gesture drawing by capturing the haphazard nature of "most mornings" and hinting to the possibility—true in this case—that perhaps I do not always have my shit together. It's not essential to the research and it does little to challenge dominant ideologies or cultivate best practices in place-based education, but if we are to seduce readers into thinking more deeply, I think it is a vital aspect of this work. There is an express desire in portraiture to reach readers beyond the academy and I contend that authentic voice—vulnerable, humourous at times, full of righteous indignation perhaps, or whatever rings most true to the lived experience—is the most effective avenue, risky as it may seem, to achieving this appeal.

One final comment on the beautiful in research. Do not forget its power, despite the fact it's not likely a topic that will be covered in most methodology courses. There is decidedly not enough space here to delve into Hans Georg Gadamer's *The Relevance of the Beautiful* (1987) in any detail, but suffice to say that hermeneutical aesthetics and portraiture share a profound respect for art's ability to reveal the limitations of fixed cultural expectancy and open the reader towards "the other." Obviously both intellectual traditions have been primarily concerned with the human "other" until recently, but "going eco" offers exciting opportunities for revitalizing portraiture by focusing on the aesthetic experience or address of non-human others. This includes arts-based explorations of place, such as those employed by Piersol in her doctoral work, but is not limited only to the arts. Aesthetic experience in this sense is understood hermeneutically, as Jardine explains: "Aesthetic experience, as Gadamer suggests, is the experience of being drawn out of our subjectivity and into a teeming world of relations that lives 'beyond our wanting and doing.' The task posed to understanding at such a juncture cannot be simply one of corralling that teeming world back into the confines of our constructs" (2006, p. 271). Aspiring ecoportraitists ought to consider employing such experiences and compositional strategies in all their possible manifestations—contemplative, seductive, shocking—as a hermeneutic means of disclosing our cultural presuppositions (the critical) and drawing us into a teeming, polyphonic world (the beautiful). What more is there to say? One day in the mountains is a thousand years.

Be patient.

Take some risks.

Pay attention.

References

Abram, D. (1996). *The spell of the sensuous: Perception and language in a more-than-human world.* New York, NY: Vintage Books.

Affifi, R. (2016). More-than-humanizing the anthropocene. *The Trumpeter, 32*(2), 155–175.

Bai, H. (2009). Reanimating the universe: Environmental education and philosophical animism. In M. McKenzie, P. Hart, H. Bai, & B. Jickling (Eds.), *Fields of green: Restorying culture, environment and education.* Cresskill, NJ: Hampton.

Barrett, M. J. (2013). Researching through porosity: An animist research methodology. In G. Harvey (Ed.), *Handbook of contemporary animism.* Durham, UK: Acumen Publishing, *20*(1), 56–69.

Blenkinsop, S. (2012). Four slogans for cultural change: An evolving place-based, imaginative, and ecological learning experience. *Journal of Moral Education, 41*(3), 353–368.

Blenkinsop, S., & Beeman, C. (2008). Dwelling telling: Literalness and ontology. *Paideusis, 17*(1), 13–24.
Blenkinsop, S., & Beeman, C. (2010). The world as co-teacher: Learning to work with a peerless colleague. *The Trumpeter: Journal of Ecosophy, 26*(3), 26–39.
Blenkinsop, S., & Fettes, M. (2009). Aligning Education and Sustainability in Maple Ridge, BC: A Study of Place-Based Ecological Schooling. Successful SSHRC eCURA Grant Application.
Blenkinsop, S., & Piersol, L. (2013). Listening to the literal: Orientations towards how nature communicates. *Phenomenology & Practice, 7*(2), 41–60.
Bringhurst, R. (2006). *The tree of meaning: Thirteen talks.* Kentville, NS: Gaspereau.
Buber, M. (1970). *I and thou* (W. Kaufmann, Ed.). New York, NY: Charles Scribner's Sons.
Clingerman, F., Treanor, B., Drenthen, M., & Utsler, D. (Eds.). (2013). *Interpreting nature: The emerging field of environmental hermeneutics.* New York, NY: Fordham University Press.
Derby, M. (2015). *Place, being, resonance: An critical ecohermeneutic approach to education.* New York, NY: Peter Lang.
Descola, P. (2013). *Beyond nature and culture.* Chicago, IL: University of Chicago Press.
Egan, K. (1997). *The educated mind: How cognitive tools shape our understanding.* Chicago, IL: University of Chicago Press.
Fettes, M. (2011). Sense and sensibilities: Educating the somatic imagination. *Journal of Curriculum Theorizing, 27*(2), 114–129.
Fettes, M. (2013). Imagination and experience: An integrative framework. *Democracy & Education, 21*(1), 1–11.
Grusin, R. (Ed.). (2015). *The non-human turn.* Minneapolis: University of Minnesota Press.
Jardine, D. (1998). *To dwell with a boundless heart: Essays in curriculum theory, hermeneutics, and the ecological imagination.* New York, NY: Counterpoints.
Jardine, D. (2006). On hermeneutics: "Over and above our wanting and doing." In K. Tobin & J. Kincheloe (Eds.), *Doing educational research: A handbook* (pp. 269–288). Rotterdam, NE: Sense.
Jardine, D. (2012). *Pedagogy left in peace.* New York, NY: Continuum.
Kohn, E. (2013). *How forests think toward an anthropology beyond the human.* Berkeley, CA: University of California Press.
Latour, B. (2013). *An inquiry into modes of existence.* Cambridge, Massachusetts: Harvard University Press.
Lawrence-Lightfoot, S., & Davis, J. H. (1997). *The art and science of portraiture.* San Francisco, CA: Jossey-Bass.
Locke, J. (1693/1996). *Some thoughts concerning education.* Indianapolis: Hackett Publishing Company.
Martusewicz, R., Edmundson, J., & Lupinacci, J. (2011). *Ecojustice education.* New York, NY: Routledge.
McKenzie, M., & Bieler, A. (2016). *Critical education and sociomaterial practice: Narration, place, and the social.* New York, NY: Peter Lang.

Piersol, L. (2015). *Eco care: Nurturing possibility & resistance within education* (unpublished doctoral thesis). Simon Fraser University, Burnaby, B.C.

Plumwood, V. (1993). *Feminism and the mastery of nature.* London, UK: Routledge.

Plumwood, V. (2002). *Environmental culture: The ecological crisis of reason.* New York, NY: Routledge.

Ranciere, J. (2004). *The politics of aesthetics.* New York, NY: Continuum International Publishing.

Rose, D. (2013). Val Plumwood's philosophical animism: Attentive interactions in the sentient world. *Environmental Humanities, 3*(1), 93–109.

Seidel, J., & Jardine, D. (2014). *Ecological pedagogy, buddhist pedagogy, hermeneutic pedagogy: Experiments in a curriculum for miracles.* New York, NY: Peter Lang.

Smith, L. (2012). *Decolonizing methodologies: Research and indigenous peoples* (2nd ed.). New York, NY: Zed Books.

Tuck, E., & McKenzie, M. (2015). *Place in research: Theory, methodology, and methods.* New York, NY: Routledge.

Tuck, E., & Yang, K. W. (2012). Decolonization is not a metaphor. *Decolonization: Indigeneity, Education and Society, 1*(1), 1–40.

Viveiros De Castro, E. (2015). *Cannibal metaphysics.* Minneapolis, MN: University of Minnesota Press.

Zwicky, J. (1992). *Lyric philosophy.* Toronto, ON: University of Toronto Press.

Zwicky, J. (2014). *Wisdom & metaphor.* Calgary, AB: Brush Education.

Zwicky, J. (2019). *Auden as philosopher: How poets think.* Nanaimo, BC: Institute for Coastal Research.

Epilogue: Humility, Resilience, Becoming: Ecoportraiture as Conversation

Sean Blenkinsop, Mark Fettes, & Laura Piersol[1]

Wrestling with Beginnings

MF: There's a lot in this book. I love the different voices in each chapter.

LP: Yes, although one thing that came up for me, reading through the whole thing, is that I'm a bit wary of anything that sounds like a recipe for "this is what it is." Each of us in our own way was trying to wrestle through just the beginnings of what it could look like.

MF: I agree that all of the statements that "ecoportraiture is this, or is that," by anyone, need to be heard as "in my experience" or "for me" or "at this moment." But the tone of the different chapters reflects people's different positionings and backgrounds. Kelly, for instance, writes like someone trained as a scientist; I like having that voice in there, even if it comes across as more definitive than other, more qualitative voices.

SB: I think the book is both these things. We can read the pieces as individual ways in which people are picking this project up and making sense of it for themselves. But the pieces also point to some general themes, for example that it couldn't be ecoportraiture if you didn't spend some time thinking about the voice of the natural world. What's not clear to me is how far we want to go. Would it be ecoportraiture if you didn't wrestle with Indigenous land-based conversations? Or if you weren't doing a particular kind of attentive listening? Those kinds of things.

[1] This Epilogue is the product of two conversations between the editors in June and September, 2020. Because of circumstances beyond our control, Laura was unable to take part in the second of these conversations, and so she is absent from some sections of the dialogue.

MF: I do feel that ecoportraiture is a bit of a process, even in the course of a single research study. That is, one learns to listen through the practice of listening. Maybe the movement towards becoming a better listener is more essential than where you start or where you end. After all, as you say yourself somewhere, Laura, and as Kelly and Michael allude to, the sheer abundance of things going on in these settings is beyond anyone's capacity to fully attend to.

LP: Yes, maybe we could talk about strivings, or directions… I like the thought of it as a process.

MF: David Jardine (2015) has this wonderful phrase about hermeneutics, that it's *educative in its intent*: "It wants to listen, to affect and to invite, not merely inform" (p. 235). Without necessarily framing itself as a learning process, that is nonetheless what ecoportraiture is. So what you're referring to as *striving*, we could say is the *process* of becoming a more ecologically educated person – a more resonant person, maybe.

LP: It's still useful to have concrete and clear things we are striving for, otherwise everything is everything.

MF: Yes, you have that little worry at the end of your piece about relativism. Presumably, this is where the search for goodness comes in? Something that came out in all the pieces, I think, is this sense of joy, or abundance, in the ordinary. Particularly the ordinary experiences of children—that was so beautifully expressed by both you and Kelly, Laura.

LB: For me, that goes back to this sense of attending or attunement. It's not just joy, it's this orientation to be present to the nuances of the more-than-human world. I struggle with the notion of goodness… I know that term is intentional, it's there to be struggled with, but still…

SB: It seems to me the goodness piece has at least two components. One is justice: a move away from deficit thinking, the way for instance that research was historically done in Black schools and Black communities. But I also really like the kind of contextual particularity to it—it's not my definition of goodness, as the researcher, or the culture's, but it's what is understood to be good, or rich, or fecund, in these particular locations. That becomes part of the challenge for the researcher, to become part of the situation to the extent that you understand what is good in this locale.

MF: But then there's also the move that Mike makes, most clearly of all the pieces, of asking, "Does what appears to be good here, in this moment, look as good if we zoom out?" You know, just beyond that line of trees is this clearcut. Or further up and further out. Are our local experiences of the good framed and supported by a larger not-good?

SB: In Mike's piece, if you make the ontological move he's proposing, if eco-portraiture research entails the researcher *being differently* in location, then what happens is that knowing and gooding changes as well.

MF: So then maybe the question of attunement that Laura is talking about is also a question of *scope*. Maybe attunement draws you into a relationality that gets vaster and vaster as it progresses. It's still particular, in a sense, but it takes in more and more of the world.

Soft Eyes and Deer Ears

SB: I've just been reading this book by Jon Young (2012), *What the Robin Knows*. He's thinking about bird languages, and the stories that are going on, and how you come to know a place. He outlines five ways in which birds communicate and how you can learn to follow the conversation, but then he also talks about the characteristics that he likes to teach students as they're becoming birders. One of them is about respect, very clearly, and how we come into spaces, including where we are emotionally—he is really convinced that if we come in angry, say, or frustrated, then the birds will pick up on that and respond. And then he uses the "soft eyes" of owls as an example of how to see as a birder. What most of us do is see in a fairly narrow tunnel of vision, which is a great cultural metaphor as well, whereas if you take on soft eyes like owls, it means you can pick up a lot more in your peripheral vision.

So all this might apply to us as researchers, how we might train ourselves to see in a much larger frame and pick up interesting things that are going on. He also talks about "deer ears," that ability to move your ears and pick up particular sounds and know what's going on, and about "working together," how birds are constantly listening not just to their own species but to their neighbours. He's clear that this is all about language, these birds have cultures, they have conversations, they have deep awarenesses and knowings about what's going on in their world.

MF: That's a really nice example of how the more-than-human world might teach us better, or perhaps fuller, ways of being human... In fact, I'd say that this is exactly what Indigenous cultures do, or did until colonization disrupted those relationships, and what many are now striving to relearn. It's part of why questions of land are so important.

SB: Yeah, and it's a question of privilege. The other day I was sitting here on the porch and a barred owl came and sat right above me, and the local robins were just freaking out, the two whose territories it was overlapping were just going crazy, and my human neighbours were walking up and down the path right underneath the owl and didn't notice. It seems like a really powerful metaphor when you think about the Black Lives Matter conversation, for example. There's a real privilege in not having to listen to what's going on in the world around you. You can just walk through, it just doesn't matter.

LP: I had this experience after living on the Prairies for a while. I had a job where I was trying to get people to connect with place and I never really thought about how it was affecting me over time. And then, at the time I was getting ready to move, I went for a walk on the prairie when the sun was shining. At that point I had this knowledge of the grasses, and the land itself, and I was hit with this awareness of having been there before and having not had that knowledge, just seeing it as an empty field... or at least with a minimal understanding, compared to what it was at that point. And with that comes the holding of all those different voices and teachings that are in that place.

MF: So let's think about that diversity for a moment. There's that sense with Kelly's piece that, the more voices you have, the truer the portrait can be. It's in that range of individual understandings, and the way they are all connected, that a fuller understanding emerges. Could we say that is generally true? That is, one can do micro-ecoportraiture, as Laura, arguably, is doing, where there's a real zooming in on individual kids (for example) and their particular understandings and interactions with the more-than-human; but if we want to understand the full complexities of the place as it stands apart from but in relationship to humans, the more voices we bring into the portrait, including, for example, Indigenous voices, the more clearly we see it. Does that sound right?

SB: Except that because this is an equity conversation, you're not just listening to *more* voices, you're listening *differently*. Historically, people may have gathered a multiplicity of voices but then they've focused on, say, the White male voice. Here, we're listening not only for the voices of different participants and actors, but also for the different voices within any particular actor. This is a piece of what Laura does so nicely with, say, Ecoboy. We actually hear these two voices going on within Ecoboy, and rather than just listening to that first, seemingly dominant voice, you also hear that other, undertone voice. This is the kind of thing Carol Gilligan does all the time too (e.g. Chu & Gilligan, 2014).

LP: For me that's one of the strivings we were talking about earlier, trying to add in and listen to all those different voices before you even begin the process of composing a portrait of your own. In my research I was trying to actually collect portraits of the students and the parents and the teachers and then myself thinking into what the more-than-human would write.

Relational Knowings

MF: In that regard, I like that you picked up on that poetic attentiveness that Don McKay (1995) writes about, and Michael picked up and amplified. That seems a really important way of understanding the listening stance, or humility, required to write a good portrait.

LP: Humility was one of those themes that came up for me. Not necessarily the ability to set aside your own voice, but to bring others into the foreground and learn from them and allow them to inform yours.

MF: That's a quality that really comes across in Robin Wall Kimmerer's (2013) writing as well. There's humility, there's reverence—that's another aspect of it that comes up again and again in these different portraits.

LP: I add in there the word *caution* as well. I think others expressed this really beautifully. It's a way of honouring the weight of the work. I was aware that I was putting these frames around understanding the experience. I think that's why I put in those ethical starting points there. In looking towards those other voices, those other stories that can contribute to the portraits, the more-than-human world is not one of those voices that we are actively listening to. I think that's an essential starting point in ecoportraiture, foregrounding those voices, and I would argue Indigenous voices too. I'm wary of coming to this work with the assumption that these landscapes are empty of cultural knowledge.

SB: And Sylvie gives us the respect, responsibility, relationship piece.

LP: For me, the opening of questions and the opening of possibility was coming from the fact that the relationships I was building through the process were pedagogical. The opening up would not have been possible had I not been listening to what those relationships were teaching me.

MF: Would you say, in a sense, that the knowing that emerges in ecoportraiture is held more in the relationships than in the researcher or in the participant—in the in-between?

SB: For me that brings up Maxine Greene's (1995) notion of the in-between... knowledge created as the result of a different relationship. She always uses that art example, of the viewer encountering a painting; both the viewer and the artist are reaching out towards one another, as it were, and it's in the space between them where the meaning of the painting emerges.

So in this case, as Laura builds relationship with Bambi, Bambi builds relationship with Laura, but they're also building relationship with place, and slug, and worm, and all those kinds of things... that knowledge is then not held by "Laura, Bambi, worm, slug" but by laurabambiwormslug. And that's what an eco-portrait might manage to convey, to some degree.

MF: Yes, I think that is what the methodology, or approach, is pointing towards: getting better at noticing how knowledge is held in these webs of relationship and getting better at describing that, or portraying that.

LP: For me, that opening up of the possible was so important. I had been reading a lot of Jacques Rancière (e.g. Bingham, Biesta, & Rancière, 2010), so all I could think of is that I was expanding my distribution of what was sensible, what was thinkable, feelable, doable.

SB: It fits nicely, because Rancière is also about the decentering of the teacher... decentering of teacher, decentering of researcher, decentering of knower. And this is where risk-taking comes in, as Mike points to. Risk in terms of what knowledge is, and risk of control, but also presumably risk to identity. Researcher changes. This may be part of that process you were talking about, Mark—working in ecoportraiture, you start to see differently, hear differently, enter particular spaces differently. All of a sudden, different voices appear, different populations feel comfortable talking to you.

MF: There's a question about the pace of this process that I've been struggling with in the context of a student's recent grant application. She's proposing to design a couple of courses, and teach them, that engage the students in a much more active and reciprocal way with the more-than-human world, and she brings in this notion of nature as co-researcher. And my question to her, or thing for her to think about, was about timescales. It's not that I am skeptical about nature as teacher or nature as co-researcher in the long haul. It just seems to me that nature's timescale is so vastly different to the scale of the semester. The teaching that nature does, and maybe most especially the co-researching that nature does... nature will hardly be getting started with you after three or four months! So how do we deal with that?

SB: I think it's a really important point. Another problem is this idea that me the human goes out and is like, "OK, let's go research, gather round!" and the natural world is like, "Oh, yeah, let's do that!" If you were doing that with humans in community research, we'd think that was weird, in and of itself, and probably inappropriate, and probably having all kinds of colonial repercussions and power repercussions.

But I've been playing with this in a different kind of direction that has to do with nature as teacher. Because I think there's a particular kind of teaching that nature can do in a short, fairly immediate timespan—"small child heads outside, falls in cold water, figures out cold and wet simultaneously"—great teaching. That kind of timing would be the easiest to work with when we start to think about nature as co-researcher—you find something immediate, or there's an immediate question that appears. But I think the natural world has taught me in these ontological kinds of ways, in interesting epistemological kinds of ways, in deep axiological kinds of ways, but that's happened over a long, long period of time. And I'm still nowhere near having achieved that or finished that. So, this question of time is profoundly important. Even just in terms of research as being, like, a DNA spontaneous gene flex and then testing it in an evolutionary context. If you follow Stephen Jay Gould's kind of conversation (e.g. 2002), that's what it's like: "Here's an accident. Let's see if it works."

MF: Yeah, let's give it 100,000 years and see what happens.

SB: So I do agree. I think what humans will do is, we'll try to circumvent that longer timescale, right, we'll go, "OK, being in the natural world takes 15 minutes a day of, like, stretching next to a tree," and what we'll lose is the profundity of this really, really extended kind of time and the learnings that arise out of that.

MF: Yeah, I definitely get that out of listening to Indigenous cultures work through these things. Elders are valued because they've been at it for so long. There's kind of an acceptance that, yes, some things just take time. You need a lifetime of listening in order to really hear well.

SB: Well, you think about the patience in response to colonialism.

MF: Oh, quite, it's unbelievable. I was listening to angel Kyodo williams in an interview yesterday (Williams, 2018), and she alluded at one point to the amazing grace of both Black American culture and Indigenous culture, this kind of cultivation of a generosity of spirit or a sort of centeredness that has managed to endure such incredible brutality over such a long period of time. Maybe this question of grace is also what ecoportraiture is aiming towards. I certainly hear it in the way Sara Lawrence Lightfoot writes (Lawrence-Lightfoot & Davis, 1997). One of the things she's looking for is this kind of resilience, or this inhabiting of a truth that endures. The goodness that she's aiming for, I think, is a kind of enduring goodness—it's something tough. It's not just aspirational, it's actually enacted continually in the face of circumstances that are restrictive and imperfect and oppressive in various ways. So maybe that's a way of thinking about the notion of goodness in portraiture that is also helpful for thinking about it in terms of the natural world. Like, what is *enduring* here?

SB: Nice. A couple of things spring to mind. One, I'm always impressed with the patience of the natural world. There seems to be a way in which it offers belonging, it offers comfort, it offers peace to everybody, no matter who you are, no matter what an idiot you are, if you'll only stop and sit. It seems incredibly generous that way. There's so much violence perpetrated on nature, and yet we still get these kinds of deep feelings of comfort and care, and that kind of peace. So I like that notion of thinking about the natural world as having this kind of enduring generosity, enduring goodness, and a tough kind of resilience to it, and that in spite of everything that's happened to us we're going to hold our integrity, we're going to hold our goodness, we're not going to descend to the level of you the colonizer. And I think that's pretty rich as a gift and as an example. I think about the Elders I spend time with—these are extraordinary human beings, that's an extraordinary example to even vaguely aspire to.

Piece number two is around the longitudinalness, if we start to think about ecoportraiture as work that extends over this much, much longer period of time, which I think is another really important offering. You can't actually do good ecoportraiture in two months, that's not how it works. So in fact, there is an ask there: to have you become part of the community, to have you be empathetic, to have you think about those kinds of relationships, to have you to, in some ways, be invited in and belong.

The Ethics of Vulnerability

LP: One of the discoveries for me over time is how this work has an ethical weight to it that is tied up in these relationships. Back when I started, I was clearest about my care for the more-than-human world; in some ways the human wasn't as interesting to me. But I was surprised at the end how attached I was to Bambi and Ecoboy and Raven, there was this complexity there to their humanness that allowed me to feel this deep attachment.

MF: You talk in your piece about empathic regard… Would you say it's a different kind of empathic regard that develops over time? That is, you get a sense of these beings, human and more-than, as dynamic and evolving, and you become, as it were, implicated in this dynamic quality?

LP: Yes, I speak to this in my thesis later on, feeling ethically answerable to them. I still feel that when I hear about how the children are doing, I still feel caring towards them.

MF: This is a kind of vulnerability, isn't it? As one gets drawn into caring for other beings, the potential for oneself being wounded increases as well.

LP: Yes, this is part of what I felt in that prairie field. Nearby they had just cleared a section of it for "development"; having become aware of how much knowledge there was in that place, I felt... *aaah!* I constantly walk that in terms of the more-than-human.

MF: Ernest Becker (1997) has that thesis around the "denial of death" driving so many pathologies of our society. It strikes me that the denial of the death of other beings, the denial of our implicatedness in the more-than-human world around us, is a piece of that pathology. So then, Sean, maybe that disregard you were talking about, the people walking by and the way we shut ourselves down, is driven by avoidance of grief.

SB: That takes us to John Bowlby (e.g. Heard, 2018) and the response to loss. You get two kinds, one is this kind of anxious attachment, the other is this overattachment, which for Gilligan scale directly on to male/female tropes within the patriarchal structure. And that patriarchal structure then gives us a psychological framework that keeps us from working through the loss to return to authentic relationship. So, yes, I think it's less about our own death, and more about the death of relationship.

LP: Sometimes it feels unethical to be connecting people to the more-than-human world right now, when so much loss is coming. I feel this with a lot of the teachers I work with—I get them to build a relationship with their micro-sites and then their micro-sites get cut down or whatever.

SB: We're setting them up for pain, quite clearly, but then we're also setting them up to be actively marginalized by the culture in which they live. In the gender conversation, the same thing's going on. It's fairly easy to get men to open up and be vulnerable, to share their desire for a different kind of relationality—but then they go to the bar. This is the educational piece of all of these conversations: how do you support people and the community into being differently in the world? How do you support nascent attempts at vulnerability, nascent attempts at connection, in a culture that spends so much time breaking connection and causing that trauma of loss? Once you've opened them up, how do you create a space that allows them to build the strength, build the fortitude, build the resilience, build the allies that would be necessary to allow them to genuinely be differently in the world?

MF: Recently, for a different project, I interviewed Cheryl Rose, a social change educator who helps run a one-month residency course in Banff for activists from around the world. She said that for her the most interesting cases are people who come out of it with a resolve to show up differently when they get back home, in the work and other social settings they are part of. They do, and it makes the people around them incredibly uncomfortable; they've found a way of succumbing less to the social pressure to dampen down their own sensibilities. What I understood her to be saying is that the showing up differently is much more pedagogical than anything else. But she was also very clear that people need support from the program in order to keep doing this—that's one of her key roles. They're not just left to sink or swim.

This came to mind because, if we're thinking about what comes next for ecoportraiture, and people read the book and feel inspired to do similar work in their own places, the question would be, can they become something of a community, an ecoportraiture community, that shares their evolving understanding, the portraits they write, but also moral and spiritual support for each other.

SB: And also academic and intellectual support. I'm sure lots of folks are still being accused of being flaky and not rigorous enough in their work.

MF: And perhaps also of not being critical enough. There's a gentleness or subtlety to portraiture; in its search for goodness and its attention to nuance, it doesn't push back in as visible a way on all the injustices and forms of oppression. For a lot of anti-racist folks, I think it's not *obviously* an anti-racist methodology.

SB: Interestingly, when you think about the resurgence conversations of Indigenous communities in Canada, it's obviously about voice, for one thing; it's also about the arts, that then allow different forms of expression and a building of pride. Some of those things are there quite clearly in ecoportraiture.

LP: I agree. Nora and I wrote a paper for one of the conferences we were at, and for me it was very clearly a political act, trying to foreground some of the voices that were not being heard. It can come across as subtle, but that is part of its power. For me, writing portraits of children definitely changed my voice around that.

But I do think there's a lot more places you could take this. I would love to see it expanding. Inviting people to take this in a more justice-oriented direction, to deepen the work...

When Voices Burst Out

MF: These days, people often express justice work in terms of being an ally. I really like something that Robin DiAngelo said in an interview that I heard a few weeks ago (DiAngelo & Menakem, 2020). The question to ask yourself is not, "Am I an ally?", because you won't be all the time. It's more, "Am I being the best ally I can be right now?" And that really resonated for me. To me it's not an identity, it's an aspiration, I guess, or something like that.

SB: I tend to think it's not an identity that we can actually claim for ourselves. It's an offering of the Other, to us. There's something artificial about saying, "Oh, I'm an ally," when it is, in fact, *their* gift that they may bestow, or their honouring, or their noticing. With the natural world, there's moments where I feel like I've managed to connect in some kinds of ways, but there are lots of moments where I see my own failings and I trip over my own feet—my anthropocentric feet, my human colonial feet... It's just constant. The response is not explicitly, "No, that's a failing," it's like, "Oh. You need to step back." But maybe it's just positing it as a possibility. We have language and conversations around allyship, so what does it mean to add that language in here, in this situation?

MF: What I feel more drawn to is the something like the notion of "the righteous Gentile" in relation to the Holocaust. Maybe this taps into the theme of humility that we were exploring. It's about ceasing to see oneself as somehow over and above, or separate from, the beings who are suffering. It's more like "What happens to the Jews happens to me," or, "What happens to the more-than-human world happens to me."

SB: A couple of things spring to mind when I start to think about that. One is Sartre's conversation (2001), where he basically posited, "I cannot be free, unless all people are free." I tend to translate that into, "I cannot be free, unless all beings are free," and wonder what that starts to look like, what the implications of that start to be. For me there's always that kind of coupling of freedom with wildness and that ability to sort of self-determine, to flourish, to mutually flourish, in a beneficial kind of way.

MF: There's something really important there, I think, around freedom, and what the portraiture literature calls voice. There's something in that striving to hear *voice*, and to portray voice in all its selfhood, that seems like a kind of championing of freedom, or maybe a singing of freedom, if you think of it as something akin to the Gospel tradition in Black American culture, let's say.

SB: Voice is so clearly one of the things that the colonizer tries to take away. If you read the feminist literature, it clearly articulates how that authentic voice gets crushed in a patriarchal society. In relationship to the natural world, that sort of crushing of voice then gets coupled with the privilege of telling the tales of the Other—summarizing the Other, shaping it such that the Other gets frozen completely: "This is the narrative, and you don't have a voice, so you can't even change it." You see it in the rising arts programs and the reclamation of languages, these are all important things that are directly connected to voice in terms of transcultural kind of conversations, and for women, trying to find ways to create educational programs that support girls to hold that authentic voice in a place that intentionally tries to crush that voice.

MF: It seems this kind of liberatory work inevitably leads to people exploring different modes of expressing voice, different ways of listening, a kind of creativity; there's always this kind of bursting out of voicing one's self and one's truths and one's aspirations in ways that haven't been done before, even though there's often an active kind of re-valorization of a past of struggle or a past that has been suppressed and occluded. Restating that is an important piece of it, but in contemporary ways, or in reimagined ways.

SB: Albert Memmi (2013) posits that part of what's going on in that anti-colonial conversation is that the folks in North Africa are actually talking to *each other*, they're no longer talking to the colonizer. Part of that conversation amongst themselves is recognizing shared experiences, shared situations of this colonial process, but also recognizing histories and kind of pushing back on particular colonial narratives, like "you didn't have governance systems, you didn't have these ways of caring for yourself that were sophisticated enough to match our better skills" or whatever. When I play that back into the natural world, you can hear those same narratives: "the natural world doesn't have a voice," "it doesn't have systems"... We actually use different languages—like if a mom gets mad at the way somebody is interacting with their kid in a playground, we talk about that as being will and anger and thoughtful response; if a mama bear gets angry at the way somebody is interacting with their cub, it's like, "well, that's instinct." We use language to separate these things.

MF. That's a really interesting observation about Memmi and about the conversation now no longer being with the colonizer but among the colonized. So how might that help us think about ecoportraiture? How can we, in the work of ecoportraiture, start to show or portray the more-than-human world talking amongst themselves, as it were?

SB: A couple of things come to mind, although I have no completeness of an answer in any way, shape or form on this one. In the way in which Mike is talking about Laura's practice, you hear this sense of a kind of depth of ongoing relationship that has been carefully built—one that's humble and caring and all those kinds of things. So you build this relationship over time. And then starting to think about oneself as a vessel, vehicle, means for translation of some of those voices—so a kind of openness. The reason Laura does this multiplicity of art stuff, this multiplicity of methods, is that I think it allows one to be made use of by the natural world in these various communicative forms. It's a similar thing with the influence of Don McKay (1995) on some of their conversations. I think he's saying there's a way in which we as humans can become—I hesitate to go as far as Teilhard de Chardin (1975) where we become the *conscience*, I don't want to do that, but—we can become a *means*. That's it. Like a particular metaphor, that will all of a sudden appear for me, or all of a sudden appear in Laura's work, or something like that. It's actually not all of a sudden appearing, it has a place, it has a history, it's an offering of some form. And you can hear it when artists, in other conversations, talk about sort of being taken over by their art. I'm not creating the sculpture, I'm just bringing forth what is there already.

MF: So the image I get from you is that it's this kind of opening up of ourselves, so that we can be more fully inhabited by the more-than-human.

SB: Yeah, I think so—a kind of porousness, a greater fluidness, like when Mike points at the ontology kind of question. You remember that quote that Raven gives, about "the words just curl into your mind"—a nine-year-old enunciating this in a brilliant sort of way. Raven also says it doesn't happen while you're thinking. So in some ways, you can't make this happen.

MF: She does sound a lot like Don McKay there. I've been reading gradually through his entire oeuvre (McKay, 2014), and he has some really nice poems about flight and the particular ways in which the structure of birds' wings works with the air. There's a real attentiveness there that is a kind of scientific thinking: "Oh, *this* is how it's managing the drag of the air!" But then he turns it into a poem.

SB: Yeah, if we start to think about that, not just as, "evolution did this lucky thing that meant birds could fly," but as an actual interaction, as a kind of knowing, as a way of encountering the world—I think that's also a different way of listening to voice.

MF: That's a good thought, that the voices in portraiture are *knowing* voices, and that's why they're there. The portraiture is celebrating those diversities of knowing.

SB: Yeah, so I was suggesting: bird knows air, wing knows air, air knows bird. Maybe you'd end up having to be a poet like Don McKay to actually play that out as you're writing the portrait.

MF: Well I think it's a really important kind of voice in the conversation, right—the poet. Just as the scientist voice, that scientific eye that Robin Wall Kimmerer (2003, 2013) for instance brings to her plants, clearly enhances the stories that she can tell, or the ways in which she can offer a portrait that is multivocal in itself, because it embodies, not just different ways of the moss or whatever speaking, but also different ways of hearing the moss speaking—there's the scientific ear, the Indigenous ear, the maternal ear, and so on.

SB: Maybe that's another piece of building skills as an eco-portraitist. You're trying to expand the ways in which you're listening, as we've talked about before, but you're also trying to expand the forms of communication, the means of communication. The potential voices that you could use yourself.

MF: There's that tension in *The Art and Science of Portraiture* between the two authors. Because Sara Lawrence-Lightfoot very much comes across as the brilliant soloist, virtuosically moving through different voices and bringing them into hearing, whereas her colleague Jessica Davis is intentionally working with a multiplicity of people, and artists in particular, or people with a kind of arts orientation, as another way of embedding multiplicity in portraiture. So I think it's kind of helpful to think of those different strategies. One can see the central task of portraiture as being a sort of project of self-discovery or self-opening or self-development, allowing oneself to become more multifaceted in the way in which one listens and how one writes and so on. But one can also think of it as a collective endeavour, where the multiplicity comes through the relationships with participants, so that everyone is a co-researcher, really.

SB: I tend to push on the co-researcher piece, just because I think it's so easy, within my own history and cultural history, to not listen to the natural world.

So finding ways to, I don't know, turn myself into a tuning fork or something like that, and allow the resonances to bounce and echo and mix or whatever.

MF: Yes, although maybe even a tuning fork is too unimodal. I'm trying to think of something like an orchestra or, you know, a mix of different instruments.

Building Possibilities for Being

SB: Everyone in the book is pointing to the way portraiture can change how we think about knowledge is held. I think the ontological turn is an interesting one as well—changing the sense of "I–us" in terms of human and in terms of identity, but also changing the beingness of that particular red cedar and that particular squirrel, that becomes a piece of the conversation. And then we see in Laura's work, quite deeply, the kind of axiological conversation, how humility and care become part of this underlying discussion. In these various ways we're playing, pointing, trying, attempting...

MF: Yes, and I think that could have value beyond the field of research. I find it intriguing that *Braiding Sweetgrass*, which we cited in the Introduction as being akin to ecoportraiture, has evoked such a strong response from people, many of whom would not see themselves as researchers... Ecoportraiture might offer people a way of enacting the kind of poetic attentiveness that Kimmerer does, in her book, in their own settings. Or I think of that German forester Peter Wohlleben and *The Hidden Life of Trees* (2016), which has been a huge success. When people buy and read that book, probably the next time they go for a walk in the woods they'll actually experience the woods a bit differently from how they would have, so that's great educational work. But imagine if they went further, encouraged by this set of ideas and practices around ecoportraiture, to actually then themselves write about *their* wood, not Peter Wohlleben's beech wood in the middle of Germany, but *their* wood, whatever it might be. Such a practice of ecoportraiture might be deeply pedagogical, for yourself, because it requires you to be more quietly and sustainedly attentive and then figure out how to put things into words, but then as you share it with others it becomes pedagogical for them as well.

To me this suggests a potential community gathered through ecoportraiture, a network not just of readers, but also of writers, extending well beyond what we normally think of research. I wonder if we could do a little thinking into what such a constellation, a network, a community of eco-portraitists might look like; how would we conceive of ourselves, how would we support one another and what we'd be trying to do as a collective as well as in our individual places and explorations.

SB: In some ways for me ecoportraiture is about potentially offering up more possibilities for how to be in the world, how to be human in the world. But simultaneously, then, if we open up more possibilities for how to be human in the world, that actually opens up space for how to be *tree* in the world, or at least how *humans* understand how to be tree in the world. For me, the larger project—ecoportraiture being a part of that kind of project—is about hearing those voices, sanctioning those voices, so there's a kind of activist piece, in terms of this increase in possibility.

I just finished writing a chapter for a phenomenological book, "Why is phenomenology important right now" kind of thing, and I realized while I was writing the chapter that Max Van Manen (e.g. 2014) and all the academic phenomenologists talk about how doing phenomenological research helps teachers to expand their practice—because, right, it makes them aware, it teaches them some skills, it opens up their sensitivity to experience, all that kind of stuff—and it's like, great, but nobody's ever said, why don't we teach kids how to do phenomenological research? Since the claim is, it's not only good for teachers in terms of their practice, but it actually changes their lives and offers them more possibilities in terms of their life... So maybe in some ways what you're talking about is, what would it mean to actually have ecoportraiture filter down as a way of making sense of the world at all levels of humanity?

MF: Yes, I think it's fascinating to think of it as not simply an academic practice.

SB: Well, you can see with all of the authors, you can hear in their work, how it's changed who they are in the world and how they make sense of their world, and how it's changed the voices they hear and how they interpret and think about them...

MF: Yes, something I particularly love about the chapters is the sense of people actually learning to inhabit themselves more authentically, or more fully, or something like that. The individuality of each author really comes through in a beautiful way, as not something static but something that has actually grown and deepened through the work that is described.

SB: In some ways we circle back to voice. We hear more of their authentic voices. There's a space for them to be differently.

MF: That seems very close to me to what an Indigenous understanding of relationality would be. That is, becoming a full human being happens through ongoing cycles of reciprocal relationship with many different kinds of beings (Stairs, 1996), most of whom are not human. *Most* of the people that you want to be in relationship with are non-human people.

SB: Maybe this is back to the tuning fork thing, the singularity of the full human being. Maybe I'm a multiplicity of beings myself. And the encounters with a diversity of other beings then allow various beings to foreground, to background, to move around, to be present, to be honoured, to be dishonoured, pushed aside, whatever, right, there's this sort of range of the process of becoming which never actually for me ends up being being. I'm just always becoming.

MF: I think there's always a tension in human cultures around stability and stasis, and movement and becoming. Both of these are important for us. It's really difficult to so fully embrace the presence of becoming that you completely let go of any sense of a stable self. This is what you spend a lifetime working towards in certain traditions, but it's clear that the reason it takes a lifetime, except for maybe a few people with an exceptional calling, is that it's so darn difficult, and it's not just difficult because of the cultures we're embedded in, it's simply inherently not something that comes easily to human beings—at least, not by the time that we've gone through childhood and come to a stage where we can start to make conscious choices about how we want to live our lives.

SB: In the modern West we tell a fairy tale of the teleological progressive life—that I get better, I get clearer, I'm always improving, I eventually reach perfection. This is a sort of religious end. And I just don't think that's right theoretically, I don't think it's right practically. This is where Camus (2012) is talking to me, he's like, well, hang on a second, we actually have to name the chaos that sits underneath this. The order is just the human desire to try and organize that, and the whole point of the absurd is that that's actually where we sit, a sort of recognition that all order that we create is incomplete and will never actually function on top of this kind of chaos. I think my life is more interesting when it's absurd and paradoxical. I love *not* coming to the solution, I think the solution is partially problematic, and you see it built right into education, but you also see it built into the very notion of what research is. Research is about coming to an answer, coming to a solution, and then taking the next step to get to the next answer and get to the next solution, written on a world that's supposed to be completely ordered... whereas

I think, at the very least for educational researchers, we know when you're working with a whole school, that's just not what's happening.

MF: It's that tension, or struggle maybe, even, between the epistemological tradition, which seeks that sort of clarity and definition, and the hermeneutic tradition, which revels in the messiness and the unfinishedness of all meaning, of all interpretation. I thoroughly agree with David Jardine (2015) that education, in the way that it actually works, is hermeneutic in its very core. This just is how it works. You can impose epistemology on that with a great deal of effort and suffering. But even then you're not really touching the essence. It's just these little islands of apparent certainty in a sea of what you're calling chaos.

SB: Right. If you play into Michel Serres' (2005) work or Bruno Latour's (2004) work, there's this sense that in fact what modernity has been trying to do is separate itself from the natural world, in essence, do that sort of control, set the structure, set what's known, do all that kind of stuff. We never actually succeeded in doing that, but we now live in a situation, with [Hurricane] Teddy crashing on the East Coast, Covid floating around, and the fires on the West Coast, where it's clear that we can't maintain that sort of myth.

MF: Yeah, we succeeded in persuading at least a portion of humanity that that's what we were doing for a while. Now, it's really interesting, this proliferation of these crazy conspiracy theories. It's like this is the last desperate effort to hang on to some meaning in the chaos. And it really feels desperate, existentially desperate.

SB: André Malraux (1979) has an interesting conversation about what happens as large civilizations collapse. He uses Spain and he uses China to show how these entities make desperate attempts to maintain the deep, deep tropes of their cultural norms, kind of what you're pointing to at this point in time. He also then writes these lovely images of, like, this desperate Spain attempting to hold control of their historical truth. And then the geese flying over top and headed for their wintering grounds in Africa... this crazy image of the natural world just being like, "whatever"...

MF: "We'll check in with you in a few more centuries and see how you're doing with that." Yeah... In that sense, ecoportraiture is very much of the now. We're articulating something that very much seems part of a response to the collapse of modernity as a set of anchors for one's life. People need to build their lives from different foundations, and maybe this is one of them.

References

Becker, E. (1997). *The denial of death.* New York, NY: Simon and Schuster.

Bingham, C., & Biesta, G. (2010). *Jacques Rancière: Education, truth, emancipation.* New York, NY: Bloomsbury.

Camus, A. (2012). *The rebel: An essay on man in revolt.* New York, NY: Vintage.

Chu, J. Y., & Gilligan, C. (2014). *When boys become boys: Development, relationships, and masculinity.* New York, NY: NYU Press.

DiAngelo, R., & Menakem, R. (2020). In conversation. Interview with Krista Tippett for the podcast On Being, https://onbeing.org/programs/robin-diangelo-and-resmaa-menakem-in-conversation/

Gould, S. J. (2002). *The structure of evolutionary theory.* Cambridge, MA: Harvard University Press.

Greene, M. (1995). *Releasing the imagination: Essays on education, the arts, and social change.* San Francisco, CA: Jossey-Bass.

Heard, D. (2018). *Attachment therapy with adolescents and adults.* New York, NY: Routledge.

Jardine, D. W. (2015). On hermeneutics: "Over and above our wanting and doing". In J. Kincheloe & K. Tobin (Eds.), *Doing educational research* (2nd ed., pp. 235–254). Leiden, NE: Brill Sense.

Kimmerer, R. W. (2003). *Gathering moss: A natural and cultural history of mosses.* Portland, OR: Oregon State University Press.

Kimmerer, R. W. (2013). *Braiding sweetgrass: Indigenous wisdom, scientific knowledge and the teachings of plants.* Minneapolis, MN: Milkweed.

Latour, B. (2004). *Politics of nature.* Cambridge, MA: Harvard University Press.

Lawrence-Lightfoot, S., & Davis, J. H. (1997). *The art and science of portraiture.* San Fransico, CA: Jossey-Bass.

Malraux, A. (1979). *Man's hope* (trans. S. Gilbert & A. Macdonald). New York, NY: Grove.

McKay, D. (1995). Baler twine: Thoughts on ravens, home and nature poetry. In T. Lilburn (Ed.), *Poetry and knowing.* Kingston, ON: Quarry Press.

McKay, D. (2014). *Angular unconformity: Collected poems 1970–2014.* Fredricton, NB: Goose Lane Editions.

Memmi, A. (2013). *The colonizer and the colonized.* New York, NY: Routledge.

Sartre, J. P. (2001). *Being and nothingness: An essay in phenomenological ontology.* New York, NY: Citadel.

Serres, M. (1995). *The natural contract.* Ann Arbor, MI: University of Michigan Press.

Stairs, A. (1996). Human development as cultural negotiation: Indigenous lessons on becoming a teacher. *The Journal of Educational Thought (JET)/Revue de la Pensée Éducative, 30*(3) (December, 1996), 219–237.

Teilhard de Chardin, P. (1975). *The phenomenon of man.* Introduction by Julian Huxley. New York, NY: Perennial Library.

Van Manen, M. (2014). *Phenomenology of practice: Meaning-giving methods in phenomenological research and writing.* Walnut Grove, CA: Left Coast Press.

Williams, A. K. (2018). The world is our field of practice. Interview with Krista Tippett for the podcast On Being, https://onbeing.org/programs/angel-kyodo-williams-the-world-is-our-field-of-practice/

Wohlleben, P. (2016). *The hidden life of trees: What they feel, how they communicate—Discoveries from a secret world.* Vancouver, BC: Greystone Books.

Young, J. (2012). *What the robin knows: How birds reveal the secrets of the natural world.* San Francisco, CA: Houghton Mifflin Harcourt.

Contributors

Sean Blenkinsop (PhD, Harvard, 2004) is a professor in the Faculty of Education at Simon Fraser University in Vancouver, Canada. His work spans educational philosophy, outdoor and experiential education, and place-based schooling. His most recent book is *Wild Pedagogies: Touchstones for Re-Negotiating Education and the Environment in the Anthropocene* (2018).

Michael Datura is a teacher, educational researcher, and occasionally a poet. He is author of *Place, Being, Resonance: A Critical Ecohermeneutic Approach to Education* and co-editor of *Wild Pedagogies: Touchstones for Re-Negotiating Education and the Environment in the Anthropocene.* He also likes birds and mushrooms.

Mark Fettes (PhD, Toronto, 2000) is an associate professor in the Faculty of Education at Simon Fraser University in Vancouver, Canada. His work focuses on the educational and cultural interplay of land, language, imagination and community. His most recent book is an Esperanto translation of Rainer Maria Rilke's poetic masterpiece, *La sonetoj al Orfeo* (2020).

Sylvie de Grandpré is an energetic and creative educator, university instructor, and PhD Candidate. She has a long career of seeking transformative ways of learning. *Ecoportraiture* is a welcome and meaningful encounter in her life. She is passionate about nature, being outdoors, art, photography, and chocolate!

David W. Jardine (now retired) had a long and influential career as a Professor in the Werklund School of Education, University of Calgary, Alberta, Canada.

He is the author and co-author of numerous books in education, over 80 articles in refereed scholarly journals, and dozens of book chapters.

Kelly Keena, (PhD, University of Colorado, 2011) has worked in formal and non-formal environmental education for more than twenty-five years. Her current work with The Morpho Institute facilitates experiences for US-based teachers in the Peruvian Amazon Rainforest where she teaches place-based pedagogies and the benefits of nature exploration for human and environmental wellbeing.

Laura Piersol (PhD, Simon Fraser University, 2015) is a Faculty Associate in the Faculty of Education at Simon Fraser University in Vancouver, Canada. She is an experienced outdoor and nature-based educator who has published widely in the fields of environmental education and educational change.

Nora Timmerman (PhD, University of British Columbia, 2013) is a teacher, parent, organizer, gardener, dancer, and desert rat who works as Senior Lecturer in Sustainable Communities at Northern Arizona University. She loves queer, liberatory politics and teaches at the intersection of activism & organizing, ecological justice, and education.

Index

F

G

J

K

L

M

N

O

P

Q

R

S

Y

Z

(RE)THINKING ENVIRONMENTAL EDUCATION

GENERAL EDITORS: CONSTANCE RUSSELL & JUSTIN DILLON

The [Re]thinking Environmental Education book series is a response to the international recognition that environmental issues have taken center stage in political and social discourse. Resolution and/or re-evaluation of the many contemporary environmental issues will require a thoughtful, informed, and well-educated citizenry. Quality environmental education does not come easily; it must be grounded in mindful practice and research excellence. This series reflects the highest quality of contemporary scholarship and, as such, is positioned at the leading edge not only of the field of environmental education, but of education generally. There are many approaches to environmental education research and delivery, each grounded in particular contexts and epistemological, ontological and axiological positions, and this series reflects that diversity.

For additional information about this series or for the submission of manuscripts, please contact:

editorial@peterlang.com

To order other books in this series, please contact our Customer Service Department:

peterlang@presswarehouse.com (within the U.S.)
orders@peterlang.com (outside the U.S.)

Or browse by series:

WWW.PETERLANG.COM

www.ingramcontent.com/pod-product-compliance
Ingram Content Group UK Ltd.
Pitfield, Milton Keynes, MK11 3LW, UK
UKHW021839100926
8556IPUK00004B/428